# Referenced Review Questions in Molecular Medicine

# Referenced Review Questions in Molecular Medicine

**Robert M. White, Sr., Ph.D., D.A.B.C.C. (CC&TC), C.H.R.M.**
Technical Director
DSI Laboratories
Fort Myers, Florida

Medical Staff
NCH Healthcare System
Naples, Florida

Courtesy Faculty
College of Arts & Sciences
Division of Math and Science
Florida Gulf Coast University
Fort Myers, Florida

**David W. Brown, Ph.D.**
Assistant Professor
College of Arts & Sciences
Division of Math and Science
Florida Gulf Coast University
Fort Myers, Florida

**Steven A. Williams, Ph.D.**
Gates Professor of Biological Sciences
Clark Science Center
Smith College
Northampton, Massachusetts

Professor of Molecular and Cellular Biology
University of Massachusetts
Amherst, Massachusetts

Director, New England Biolabs Molecular Biology Summer Workshops
Clark Science Center
Smith College
Northampton, Massachusetts

AACC Press

2101 L Street, NW, Suite 202
Washington, DC 20037-1558

1 2 3 4 5 6 7 8 9 0 PCP 07 06 05

Printed in the United States of America

**Library of Congress Cataloging-in-Publication Data**

White, Robert M.
Referenced review questions in molecular medicine / Robert M. White, Sr., David W. Brown, Steven A. Williams.
p. cm.
Includes bibliographical references.
ISBN 1-59425-033-2 (alk. paper)
1. Molecular diagnosis—Examinations, questions, etc. 2. Molecular biology—Examinations, questions, etc. I. Brown, David W. II. Williams, Steven A. III. Title.
[DNLM: 1. Molecular Diagnostic Techniques—Examination Questions. 2. Molecular Biology—Examination Questions. QY 18.2 W587r 2005]

RB43.7.W45 2005
611′.01816—dc22 2005010272

# Contents


| | |
|---|---|
| Preface | *vii* |
| Abbreviations | *ix* |
| Questions in Molecular Medicine | 1 |
| Answers and Explanations | 143 |
| References | 221 |

# Preface

Molecular medicine is one of the most rapidly expanding fields in science today. It isn't just moving at a fast linear pace; it is growing exponentially. Thus, this book intends only a brief survey of the field.

The questions for *Referenced Review Questions in Molecular Medicine* were taken from a number of current references, including some from the authors' experiences. However, even though a number of references were used, the book cannot possibly cover all aspects of molecular medicine.

We hope that the questions presented will provide an intermediate-level challenge to the reader and encourage him or her to dig ever deeper into a fascinating field that is incredibly rich in detail.

# Abbreviations

Many of the abbreviations are defined within the question where they are used. However, a list is given below for those that are not defined. Also, some proprietary buffers and other solutions (contents unknown) are listed only by their abbreviation, as that is how they are referenced by the manufacturer.

| | |
|---|---|
| A | Adenine |
| $A_{260}$ | absorbance at 260 nm |
| $A_{280}$ | absorbance at 280 nm |
| ABL | Abelson |
| AIDS | acquired immune deficiency syndrome |
| ARMS | amplification refractory mutation system |
| ASO | allele-specific oligonucleotide |
| BAC | bacterial artificial chromosome |
| BCR | break-point cluster region |
| BRCA | breast cancer susceptibility gene |
| C | cytosine |
| cDNA | complementary DNA |
| cM | centimorgan |
| dATP | deoxyadenosine triphosphate |
| $ddH_2O$ | doubly distilled water |
| ddNTP | dideoxynucleoside triphosphate |
| del | deletion |
| DEPC | diethylpyrocarbamate |
| DNA | deoxyribonucleic acid |
| dNTP | deoxynucleoside triphosphate |
| dsDNA | double-stranded DNA |
| dsRNA | double-stranded RNA |
| dUPT | deoxyuridine triphosphate |
| EDTA | ethylenediminetetracetic acid |
| FISH | fluorescent *in situ* hybridization |
| FITC | fluorescein isothiocyante |
| HCV | hepatitis C |
| HEPES | N-2-hydroxyethylpiperazine-N'-2-ethansulfonate |
| g | gravity |
| G | guanine |
| HPV | human papillovirus |
| inv | inversion |
| kb | kilobases |
| kDa | kilodalton |
| $LN_2$ | liquid nitrogen |

| | |
|---|---|
| M | molar |
| MOPS | 3-(N-morpholino) propane sulfonate |
| mRNA | messenger ribonucleic acid |
| *myc* | avian myelocytomatosis virus |
| NASBA | nucleic acid sequence-based analysis |
| NF | neurofibromatosis gene |
| p | petit or short arm of a chromosome |
| PCI | phenol-chloroform-*iso*-amyl alcohol (50:49:1) |
| PCR | polymerase chain reaction |
| PML | promyelocytic leukemia |
| q | queue or long arm of a chromosome |
| RAR | retinoic acid receptor |
| RB | retinoblastoma gene |
| rRNA | ribosomal RNA |
| RNA | ribonucleic acid |
| ssDNA | single-stranded DNA |
| ssRNA | single-stranded RNA |
| RT-PCR | reverse transcriptase PCR |
| S | Svedberg |
| SDA | strand displacement amplification |
| SDS | sodium dodecylsulfate |
| SNP | single nucleotide polymorphism |
| t | translocation |
| T | thymine |
| *Taq* | *Thermus aquaticus* |
| TBE | Tris/borate/EDTA |
| $T_m$ | melting temperature |
| TMA | transcription-mediated analysis |
| TNF | tumor necrosing factor |
| Tris | *tris*(hydroxymethyl)aminomethane |
| tRNA | transfer RNA |
| μg/mL | micrograms per milliliter |
| UTR | untranslated region |
| UV | ultraviolet |
| VNTR | variable number of tandem repeats |
| wt | wild-type |

# Questions in Molecular Medicine

1. What is the process called in which a new molecule of DNA is synthesized using an existing molecule of DNA as a template?

   a. transcription
   b. translation
   c. replication
   d. recombination

2. What is the process called in which a new molecule of RNA is synthesized using an existing molecule of DNA as a template?

   a. transcription
   b. translation
   c. replication
   d. recombination

3. Where is eukaryotic genomic DNA located?

   a. in the mitochondria
   b. in the chloroplasts
   c. in the cytoplasm
   d. in the nucleus

4. Which of the following is a restriction enzyme recognition sequence?

   a. GATTAG
   b. GAAAAG
   c. GGATCC
   d. GAGAGA

5. What is a noncoding gene sequence that is spliced out of a newly synthesized pre-mRNA prior to export and translation of the processed mRNA?

   a. intron
   b. exon
   c. poly-A tail
   d. 5′-cap

6. Which enzyme is used in the process of eukaryotic mRNA production?

   a. reverse transcriptase
   b. RNA polymerase II
   c. T4 DNA ligase
   d. DNA polymerase

7. RNA structure is normally which of the following?

   a. double-stranded
   b. single-stranded
   c. single-stranded with some double-stranded regions due to the single-stranded RNA folding back and pairing with itself
   d. none of the above

8. In what form is prokaryotic genomic DNA?

   a. a single closed circular chromosome
   b. multiple plasmids
   c. a single linear chromosome
   d. multiple linear chromosomes

9. In the isolation of genomic DNA from cellular material, what is the purpose of the enzyme, <u>Proteinase K</u>?

   a. to inhibit endogenous nucleases
   b. to digest cellular proteins
   c. to digest contaminating RNA
   d. to precipitate the DNA

10. Which of the following specimen(s) is/are *preferred* for the isolation of human genomic DNA from whole blood?

    a. EDTA anticoagulated blood
    b. heparin anticoagulated blood
    c. clotted blood
    d. both *a* and *b*

11. Which of the following reagents is/are necessary for the precipitation of genomic DNA?

    a. sodium acetate, sodium chloride, or ammonium acetate
    b. ethanol or isopropanol
    c. chloroform
    d. phenol
    e. both *a* and *b*

12. You have prepared a sample of genomic DNA and have characterized it with UV spectroscopy. You have obtained an $A_{260}/A_{280}$ ratio of 1.77. What does this likely indicate?

    a. The DNA is of high quality.
    b. The DNA preparation contains significant protein contamination.
    c. The DNA is of high molecular weight.
    d. The total concentration of DNA in the sample is 98.5 μg/mL.

13. Which of the following statements relate(s) to plasmid DNA?

    a. small, circular DNA molecules
    b. contain gene sequences that are beneficial to the host, but not essential
    c. may contain multiple cloning sites to facilitate genetic manipulation
    d. all of the above

14. Considering tRNA, which of the following is the *true* statement?

    a. In the charged form, the amino acid is attached to the 5′-OH.
    b. The terminal CCA-3′ is coded for in the tRNA gene.
    c. Base pairing in double-stranded regions is exclusively Watson-Crick style.
    d. Class 1 tRNA molecules have a short extra arm consisting of three to five bases.

15. Which feature of eukaryotic mRNA is often exploited in its selective isolation?

    a. the 5′-cap
    b. the poly-A tail
    c. the intron-exon splice junction
    d. the promoter sequence

16. You plan to isolate total RNA from whole blood, and need to prepare a Tris buffer. Which of the following would you do?

    a. Use DEPC-treated water to prepare the buffer.
    b. Prepare the buffer and then DEPC treat it.
    c. Use tap water.
    d. Use deionized $H_2O$.

17. Which of the following <u>may</u> be used in the preparation of total RNA from cells?

    a. chaotropic agents such as guanidinium thiocyanate
    b. PCI extraction
    c. CsCl gradient ultracentrifugation
    d. all of the above

18. Which type of enzyme would you use to cleave a dsDNA at a specific site?

    a. reverse transcriptase
    b. exonuclease
    c. restriction endonuclease
    d. phosphatase

19. Which enzyme would you use to join together two separate DNA fragments?

    a. DNA polymerase
    b. restriction endonuclease
    c. DNA ligase
    d. phosphatase

**The following diagram applies to questions 20–23:**

Given the following DNA sequence, three potential restriction enzyme cleavage sites are shown:

```
           1  2  3
           ↓  ↓  ↓
5′-n n n G G A T C C n n n-3′
3′-n n n C C T A G G n n n-5′
           ↑  ↑  ↑
           3  2  1
```

20. Which cleavage site would yield blunt-ended fragments?

a. site 1
b. site 2
c. site 3
d. none of the above

21. Which cleavage site would yield a 5′-protruding end?

a. site 1
b. site 2
c. site 3
d. none of the above

22. Which enzyme would you use to convert the product of site 3 cleavage to a blunt end?

a. None; cleavage here produces blunt ends already.
b. Klenow fragment of DNA polymerase I
c. exonuclease III
d. alkaline phosphatase

23. Assume that the above sequence is present in both plasmid and genomic DNA. You cleave both these DNAs at site 1 and mix the products together. Which enzyme would you use to prepare recombinant plasmid DNA?

a. T4 polynucleotide kinase
b. alkaline phosphatase
c. T7 RNA polymerase
d. T4 DNA ligase

24. Which of the following DNA sequencing methods uses chemical cleavage of the DNA?

a. Edman degradation
b. Sanger or dideoxy method
c. Maxam-Gilbert method
d. none of the above

25. In chain termination sequencing, what is the ratio of ddNTPs to normal dNTPs typically added to the reaction mixture?

    a. 1:1
    b. more ddNTPs than dNTPs
    c. more dNTPs than ddNTPs
    d. 100:1

26. In PCR, why is the first step an incubation at 94 °C?

    a. The *Taq* polymerase is more active at this elevated temperature.
    b. Primers anneal to the DNA faster at this elevated temperature.
    c. to ensure complete melting of the dsDNA
    d. to prevent mis-priming of the 5′-ends of the primers

27. The second step of PCR is typically near 50–65 °C, but can vary. Why?

    a. The *Taq* polymerase is active over a wide range of temperatures.
    b. to avoid mis-priming the 3′-ends of the primers
    c. to prevent melting of the newly synthesized DNA
    d. Optimum primer annealing temperature is sequence-specific but typically falls within this range.

28. In designing primers for PCR, which of the following characteristics should *not* be chosen?

    a. 18–30 nucleotides in length
    b. no runs of poly-G or poly-C
    c. The primers should not be complementary to one another.
    d. The 3′-end should contain a large number of G/C residues.

29. How is the concentration of oligonucleotide primer DNA determined?

    a. $A_{260}$ using a sequence-specific relationship
    b. $A_{260}$ using the relationship of 1.0 $A_{260}$ = 50 μg/mL
    c. $A_{260}/A_{280}$ ratio
    d. dye binding and colorimetric determination

30. What is the source of the *Taq* polymerase used in PCR?

    a. *Thermus aquaticus*
    b. *Tetrahymena pneumophila*
    c. *Arabidopsis thaliana*
    d. *E. coli*

31. How many gene-specific primers are used in the ligase chain reaction (LCR)?

    a. one
    b. two
    c. three
    d. four

32. You have set up an LCR, but neglected to add dNTPs. What would be the most likely result?

    a. mis-priming at alternate sites, causing a false-positive reaction
    b. mis-priming at alternate sites, causing a false-negative reaction
    c. no extension of the primers due to lack of substrate molecules
    d. no effect; LCR does not need dNTPs

33. What is the main difference between TMA and NASBA?

    a. TMA uses two enzymes, while NASBA uses three.
    b. TMA uses three enzymes, while NASBA uses two.
    c. TMA uses thermostable DNA ligase, while NASBA uses thermolabile DNA ligase.
    d. There is no difference; both methods are exactly the same.

34. In a nucleotide triphosphate, the three different phosphate groups are denoted with Greek letters: $\alpha$, $\beta$, and $\gamma$. With respect to the ribose sugar, which position is the $\alpha$ position?

    a. the one nearest the ribose (first)
    b. the one in the middle (second)
    c. the one farthest from the ribose (last)
    d. the one on the 3′ end

35. Which of the following methods uses restriction enzyme cleavage and the repair activity of DNA polymerase to amplify DNA:

    a. NASBA
    b. SDA
    c. RT-PCR
    d. branched DNA detection

36. At the end of 25 cycles of PCR, how many DNA molecules are theoretically present in the reaction if one starts with one double-stranded DNA molecule as the template?

    a. 25
    b. 50
    c. 100
    d. $2^{25}$

37. Which of the following methods is/are used to evaluate gene expression?

    a. Northern blotting
    b. ribonuclease-protection assay
    c. RT-PCR
    d. all of the above

38. Which of the following methods is most <u>sensitive</u> for determining gene expression?

    a. Northern blotting
    b. nuclease protection assay
    c. RT-PCR
    d. SDA

39. Using gene-specific intron-spanning primers you perform both PCR and RT-PCR on samples of genomic DNA and total RNA, respectively. Following electrophoresis you note that the RT-PCR product is shorter than the PCR product. What is the appropriate action?

   a. Repeat the test; you added the wrong primers.
   b. Repeat the test; you have genomic DNA contamination in your RNA prep.
   c. Repeat the test; you have RNA contamination in your DNA prep.
   d. No further action is required; these are expected results due to intron-exon splicing of the mRNA.

40. What is the first step in RT-PCR?

   a. synthesis of A-DNA
   b. synthesis of B-DNA
   c. synthesis of cDNA
   d. synthesis of Z-DNA

41. What is RNase H?

   a. an enzyme that selectively degrades ssRNA
   b. an enzyme that selectively degrades RNA : DNA duplexes
   c. an enzyme that selectively degrades dsRNA
   d. an enzyme that synthesizes DNA from an RNA template

42. If you wanted to selectively amplify <u>all</u> of the mRNA present in a sample of total RNA, which primer would you use for the first step in RT-PCR?

   a. random hexamers
   b. oligo-d$T_{15}$
   c. gene-specific primer
   d. none of the above

43. Why is RNA first heated to 70 °C for 10 min prior to RT-PCR?

   a. to inactivate RNases
   b. to allow for primer annealing
   c. to reduce RNA secondary structure that may inhibit the reverse transcriptase
   d. to inhibit the action of residual proteinase K from the isolation procedure

44. You perform a PCR using a set of primers (both are 20-mers) that you have designed yourself using gene sequence data. Electrophoresis shows not the predicted 750 bp product, but a large amount of a 30 bp product. What happened?

   a. You used the wrong extension temperature.
   b. You used the wrong annealing temperature.
   c. You used the wrong gene sequence in your primer analysis.
   d. Your primers have a 10-base homology at the 3′ end, causing them to anneal together instead of with the genomic DNA (primer-dimer formation).

45. You wish to perform a restriction digestion of genomic DNA using an enzyme that recognizes the sequence TAGCTA. If this genomic DNA had a completely random sequence and 50% GC base pairs, approximately how frequently would this enzyme cut the DNA?

    a. every 24 bp ($6 \times 4$)
    b. every 1296 bp ($6^4$)
    c. every 4096 bp ($4^6$)
    d. every 17,280 bp ($6! \times 4!$)

46. You wish to evaluate the restriction digestion performed in question #45 by gel electrophoresis. Which type of gel would you use?

    a. denaturing 8% polyacrylamide
    b. 4% agarose
    c. 6% acrylamide under native conditions
    d. 0.8–1.0% agarose

47. In this gel image, which lane represents the restriction digestion performed in question #45?

    a. lane 1
    b. lane 2
    c. lane 3
    d. lane 4

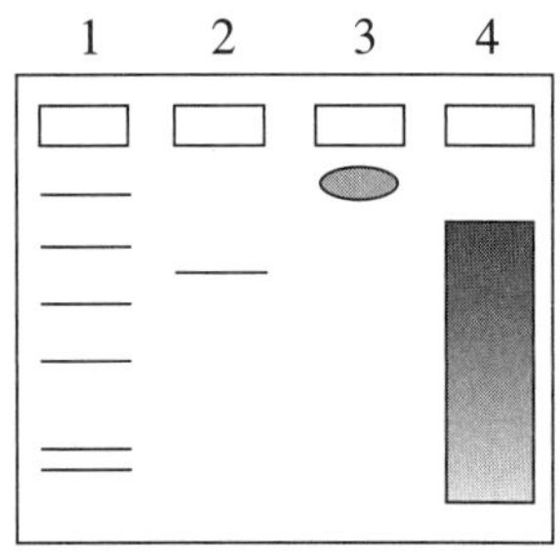


48. In this gel, <u>lane M</u> is λ/HindIII marker fragments. Band sizes from top to bottom are as follows: 23.1 kb, 9.4 kb, 6.6 kb, 4.4 kb, 2.3 kb, 2.0 kb, 564 bp. In <u>lane 2</u>, from top to bottom, what are the apparent sizes of the bands indicated by the arrows?

    a. 11.2 kb, 4.8 kb, 3.6 kb
    b. 3.6 kb, 4.8 kb, 11.2 kb
    c. 20.0 kb, 5.9 kb, 2.4 kb
    d. 9.4 kb, 4.4 kb, 2.3 kb

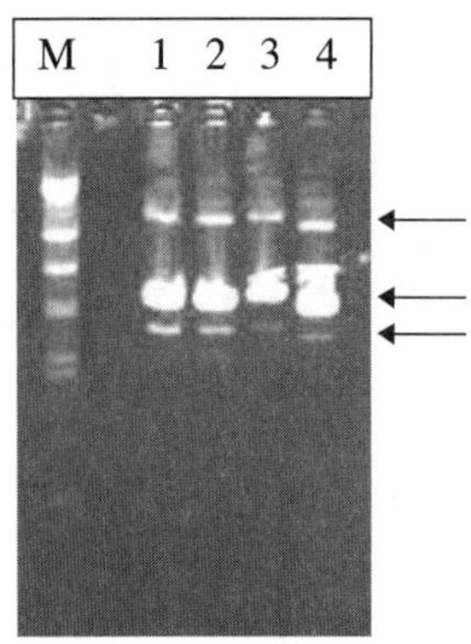

49. In the gel shown in question #48, assume that the MW markers were ***not*** boiled prior to electrophoresis and predict how lane M would appear.

   a. The 9.4 kb band would appear as a doublet.
   b. The 2.3 kb and the 2.0 kb bands would appear as a single band.
   c. The 4.4 kb band would appear diminished in intensity.
   d. There would be no observable change in appearance.

50. In the gel shown in question 48, the top arrow represents relaxed circular plasmid with a single-strand nick, the middle arrow represents tightly supercoiled circular plasmid, and the bottom arrow represents plasmid with a double-stranded break (linear plasmid). You take a sample of this plasmid and restriction digest with *Bam*H I, which has a single recognition site in the plasmid. Following this restriction digestion you run an aliquot on an agarose gel alongside a sample of the uncut plasmid. In the accompanying diagram, where would the cut sample migrate?

   a. position A
   b. position B
   c. position C
   d. position D

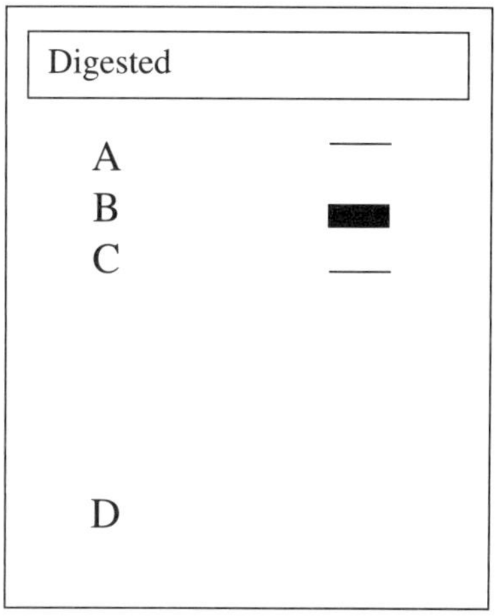


51. What is the material most commonly used for nucleic acid hybridization membranes today?

   a. nitrocellulose
   b. PVDF
   c. nylon
   d. rayon

52. In a hybridization reaction, a small amount of nonfat dried milk is often added to the solution. Why?

   a. Nonfat dried milk acts as a blocking agent to reduce nonspecific hybridization of the probe DNA to the membrane.
   b. Nonfat dried milk acts to lower the hybridization temperature of the DNA : DNA duplex, permitting more stringent washing conditions to be employed.
   c. Nonfat dried milk acts to occupy solution volume, thus artificially increasing the probe concentration in the hybridization solution.
   d. Nonfat dried milk acts in conjunction with formamide to stabilize the formation of hydrogen bonding between the probe and the target molecule.

53. What is the $T_m$ of a DNA:DNA duplex?

a. the temperature at which all of the probe and target strands are dissociated
b. the temperature at which the probe denatures into individual oligonucleotides
c. the temperature at which the bound target molecules dissociate from the membrane
d. the temperature corresponding to the mid-point in the observed transition from double-stranded to single-stranded forms

54. Fixation of transferred or blotted deoxyribonucleic acid to a nylon membrane may be accomplished by which of the following techniques?

a. baking at 80 °C for 2 h
b. exposing briefly to ultraviolet light
c. transferring with an alkaline solution to a negatively charged membrane
d. all of the above

**55–64. Match the nucleotide structures to their names.**

55. _____

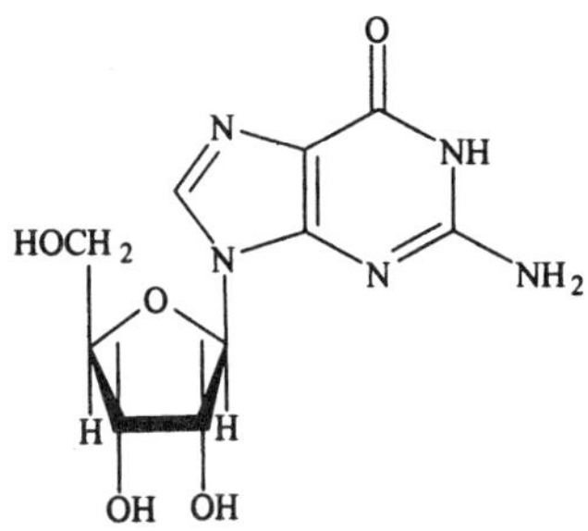


a. adenosine
b. thymidine
c. cytidine
d. pseudouridine
e. guanosine
f. dihydrouridine
g. uridine
h. inosine
i. 1-methylguanosine
j. methylinosine
k. isopentenylguanosine
l. pseudothymidine

56. _____

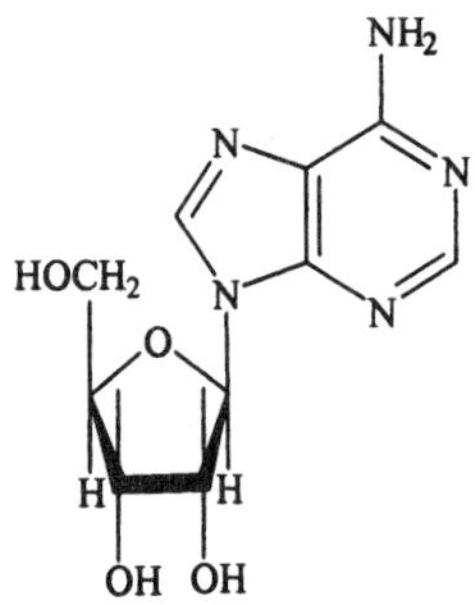

57. _____

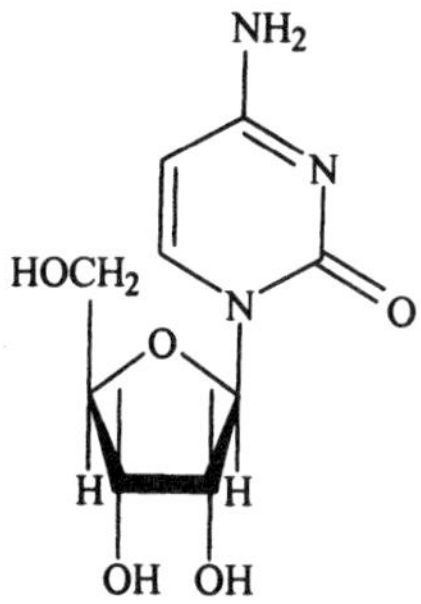

58. _____

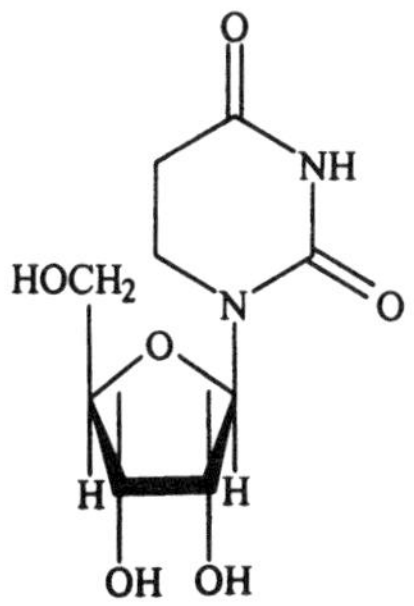

59. _____

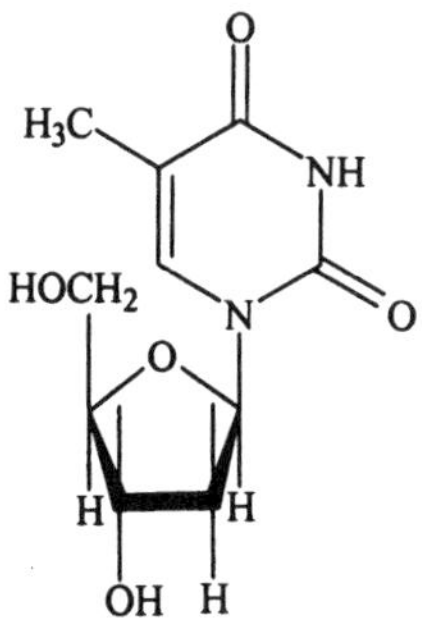

60. _____

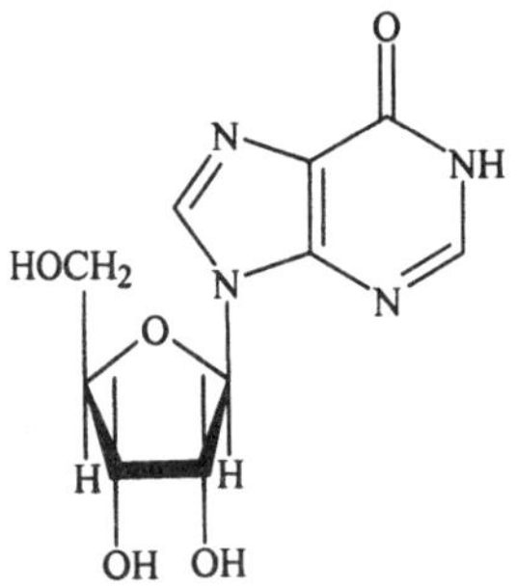

61. _____

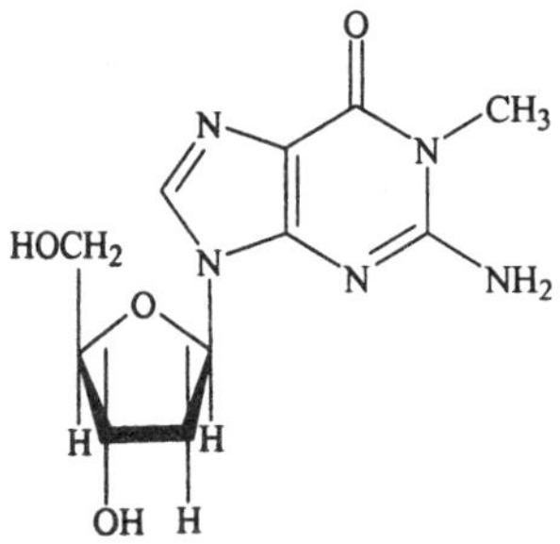

a. adenosine
b. thymidine
c. cytidine
d. pseudouridine
e. guanosine
f. dihydrouridine
g. uridine
h. inosine
i. 1-methylguanosine
j. methylinosine
k. isopentenylguanosine
l. pseudothymidine

62. _____

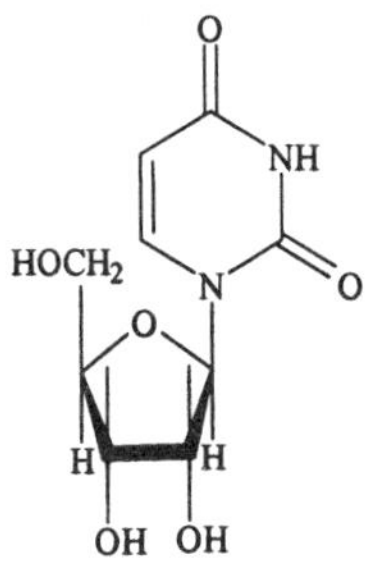

63. ______

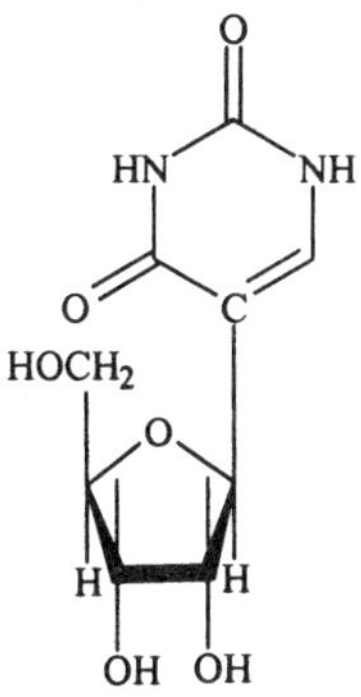


64. _____

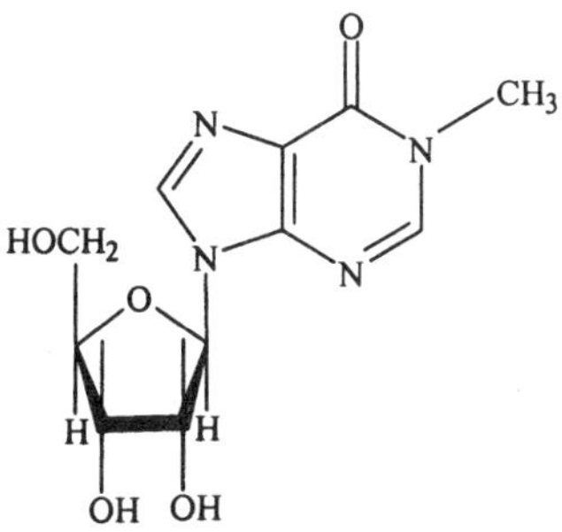


65. Which of the following is *true* according to Knudson's "Two-hit" hypothesis?

a. Both mutagenic events necessary for the development of retinoblastoma and, perhaps, other tumors, arise from germ-line events that occur simultaneously.
b. The first mutagenic "hit" may arise from the germ line and the second "hit" from a second, somatic event.
c. Both mutagenic events arise from somatic events that occur simultaneously.
d. Both mutagenic events arise from somatic events that occur in stepwise fashion.
e. Inherited tumors are expressed later in life due to the need for two identical alleles to be mutated before tumorigenesis can occur.

66. Which of the following is/are *true* about the zinc finger motif?

a. It can have zinc coordinated/bound to four linked cysteines such as might be found in the steroid/thyroid hormone receptor family.
b. It can have zinc coordinated/bound to two cysteines and two histidines such as might be found in TFIIIA, kruppel, Sp1, etc.
c. It seems to act by binding to DNA using a series of three amino acids specific for an essentially unique DNA address.
d. only *a* and *b*
e. only *b* and *c*
f. *a*, *b*, and *c* are true

67. Which of the following cellular proto-oncogenes might be amplified in small-cell lung cancer?

a. c-*myc*
b. L-*myc*
c. N-*myc*
d. all of the above
e. only *b* and *c*

**68–89**. The goal of the following experiment is to purify mRNA from mouse liver and then perform experiments to study the expression of a specific gene (*Pal*) (Northern blot, RT-PCR, and quantitative RT-PCR).

Step 1. Dissected mouse livers were cut in half (approximately 1 g per 1/2 liver) and flash-frozen in an RNase-free conical tube containing liquid nitrogen ($LN_2$). To each half-liver, we added 15 mL of buffer RLT (containing β-mercaptoethanol and guanidinium isothiocyanate), and then the half liver was homogeinized. The homogenate was centrifuged at 3000–5000 × g for 10 min.

68. Which of the following statements is *true* about guanidinium isothiocyanate?

a. It acts as a powerful inhibitor of DNase.
b. It acts as a powerful inhibitor of RNase.
c. It precipitates protein.
d. It converts dsDNA into ssDNA.
e. It converts ssDNA into dsDNA.

Step 2. Without carrying over any of the upper fatty layer, the solution is transferred over and 70% ethanol (made with DEPC-treated water) is added. The solution is shaken vigorously.

Step 3. A 15-mL aliquot of the mouse liver solution from Step 2, above, is added to a Qiagen RNA column (silica) and the column is centrifuged at 3000–5000 × g for 5 min. The flow-through is discarded, the remaining solution is added to the column, and the column is centrifuged an additional 5 min at 3000–5000 × g.

Step 4. Next, 15 mL of Buffer RW1 is pipetted onto the column and centrifuged at 3000–5000 × g for 5 min. The flow-through is discarded.

Step 5. Then, 10 mL of wash buffer RPE (concentrate diluted 1:5 with 100% ethanol) is pipetted. The column is centrifuged for 2 min at 3000–5000 × g. The flow-through is discarded. This step is repeated with centrifugation for 10 min to get rid of all excess ethanol.

Step 6. The column is then transfered to another RNase-free tube, and 1.2 mL of RNase-free water is added and centrifuged at 3000–5000 × g for 3 min. The column eluate is saved. This step is then repeated.

69. True or false: In the absence of ethanol, water is capable of eluting DNA from the Qiagen column.

Step 7. A small aliquot of the eluate from step 6, above, is diluted with DEPC-treated water. The spectrum is scanned, using a 1-cm cuvette, from 260 nm to 280 nm. In the case of this experiment, an $OD_{260} = 1.10$ was obtained. The $OD_{260}/ OD_{280}$ was 2.01.

70. Assuming that an $OD_{260} = 1$ is equivalent to the presence of 40 μg/mL of RNA, calculate the concentration of RNA present in this experiment.

Step 8. Next, 1 μL of the 200 μL of RNA solution is transferred to a separate tube and frozen at –70 °C (PCR RNA). This step is repeated (Total RNA).

Step 9. To the remaining 198 μL of the RNA solution is added 52 μL of DEPC-treated water and 250 μL of binding solution, which is then mixed.

Step 10. The RNA solution from step 9, above, is transferred to a tube of oligo dT cellulose and mixed by inversion. The RNA/oligo dT mixture is incubated at 70 °C for 5 min, and then rocked for 45 min at room temperature.

71. To what does the oligo dT mixture bind?

a. the poly-A tail on the RNA
b. wherever there is an "A" in the RNA
c. wherever there is an "A" in the extracted DNA
d. random areas of RNA that have a few "A's"
e. None of the above, since the RNA is bound at a later step.

Step 11. The RNA/oligo dT cellulose is "pelleted" by centrifuging for 3 min at 3000 × g and the supernatant is removed.

72. What is the purpose of the cellulose?

a. to buffer the solution
b. to adsorb small, colored organic molecules
c. to provide enough mass for the oligo dT, which it is bound to the mRNA, to sediment upon centrifugation
d. to adsorb the RNA
e. to adsorb any leftover DNA

Step 12. The oligo dT cellulose is resuspended in 500 μL of wash solution 1 by vortexing. The cellulose is then transferred to an RNA spin column. The column is centrifuged at 3000 × g for 3 min at room temperature and the flow-through is discarded.

Step 13. Next, 500 μL of wash solution 1 is added to the oligo dT cellulose and vortexed. The column is centrifuged at 3000 × g for 3 min at room temperature and the flow-through is discarded.

Step 14. Then, 500 μL of Wash Solution 2 is added to the oligo dT cellulose and mixed by vortexing. The column is centrifuged at 3000 × g for 3 min and the flow-through is discarded. Step 13, above, is repeated with 500 μL of fresh wash solution.

Step 15. The RNA column is transferred to a new collection tube and 100 μL of warm (70 °C) elution buffer is added. The solution is vortexed and immediately centrifuged at 4900 × g for 2 min. The eluate is collected. Another 100 μL of warm elution buffer is added, the mix is re-centrifuged, and the eluate is collected. The column is discarded.

73. What does the elution buffer do?

a. It degrades the cellulose enzymatically and allows the mRNA to pass through the column.
b. It breaks the A–T bonds between the mRNA and the oligo dT, freeing the mRNA and allowing it to pass through the column.
c. It dissolves the cellulose and allows the entire solution to pass through the column.
d. It enzymatically breaks off the poly-A tail, allowing the original mRNA to pass through the column.
e. It melts the G–C bonds, freeing the mRNA and allowing it to pass through the column.

Step 16. Now, 50 μL of the A+ RNA is transferred to a tube. Three substances are added: 5 μL of 5 M ammonium acetate, 0.5 μL of glycogen, and 140 μL of ice-cold 100% ethanol. The solution is stored at –70 °C.

Step 17. The remaining 150 μL of the A+ RNA is transferred to a tube. The following substances are added: 5 μL of 5 M ammonium acetate, 0.5 μL of glycogen, and 140 μL of ice-cold 100% ethanol. The solution is stored at –70 °C to create cDNA in a later procedure.

Step 18. To the RT PCR RNA tube from step 8, above, is added 49 μL of RT/PCR buffer. The solution is buffered and gently mixed. The RT/PCR buffer contains the following:

1X PCR Buffer
1.75 mM $MgCl_2$
0.8 mM each dNTP
10 units of RNase inhibitor
5 mM DTT (dithiothreitol)
0.4 pM/μL of the *Pal* forward primer (upstream specific)
1.4 pM/μL of the *Pal* reverse primer (downstream specific)
2.5 Units *Taq* Gold polymerase
15 Units Multiscribe reverse transcriptase

The following reactions are also set up at the same time:

A "No Template" control in which 1 μL of $H_2O$ is added instead of template.
A "No RT" control in which 1 μL of $H_2O$ is added instead of reverse transcriptase.
A "DNA Contamination" control reaction that is spiked with 75 ng of genomic mouse DNA.

A positive control reaction that includes a template that you know will amplify with the same primers used in the RT-PCR reaction.

74. True or false: Dithiothreitol also inhibits RNase.

Step 19. The tube from step 18, above, is placed in the PCR instrument. The following parameters will be used:

25 °C for 10 min
42 °C for 12 min
95 °C for 10 min
43 cycles of 94 °C for 20 sec and 62 °C for 1 min
72 °C for 7 min
Hold at 4 °C

75. What is the first step in the above PCR?

a. to heat-kill any RNase that may be present
b. to denature the dsRNA
c. to activate the *Taq* Gold
d. to anneal the *Pal* reverse (downstream-specific) primer
e. to denature any dsDNA that may be present

76. What is the second step in the above PCR?

a. to heat-kill any RNase that may be present
b. to denature the dsRNA
c. to activate the Taq Gold
d. to anneal the *Pal* primers
e. to synthesize the first strand of DNA

77. What is the third step in the above PCR?

a. to denature the cDNA/RNA complex
b. to heat-kill the reverse transcriptase
c. to activate the Taq Gold
d. all of the above
e. only *a* and *c*

Step 20. A 1.8% agarose gel is poured.

Step 21. Then 8 μL is removed from the "RNA PCR" tube, placed in a new tube, and mixed with 2 μL of "Blue Juice" (20% ficoll; 0.01 M Tris-HCl, pH 8.0; 1 mmol $Na_2$-EDTA; and 0.1% xylene cyanol FF 0.1% bromphenol blue).

Step 22. The gel from step 21, above, is loaded in the following order:

10 μL 100 bp ladder in "Blue Juice"
10 μL "RNA PCR" from step 21, above
10 μL pBR322 – *Msp*I Digest in "Blue Juice"
18 μL "No Template" control
18 μL "No RT" control
18 μL "DNA Contamination" control

Electrophorese at 70 VDC for 1–2 h. Stain (ethidium bromide), destain, and photograph.

78. True or false: The pBR322 *Msp*I digest provides a series of approximately 26 fragments that range in molecular weight from 110 to 622 bp.

79. True or false: The intensity of the bands in the 100 bp ladder may be used as an estimate of the amount of DNA present in the "RNA PCR" by comparing the intensity of fluorescence in a band close in molecular weight and of known concentration to a marker band that is in the unknown.

Step 23. The RNA from step 16, above, is recovered by centrifuging for 60 min at 4 °C. The supernatant is removed and then 1 mL of ice-cold 70% ethanol is added. The solution is vortexed for 30 sec and then centrifuged for 15 min. Again, the supernatant is removed and re-centrifuged, and excess ethanol is removed. The RNA pellet is allowed to air dry for 10 min.

Step 24. The following substances are added to the tube from step 23, above:

15 μL of Northern Loading Dye
720 μL deionized formamide
160 μL 10X MOPS, which is 3-(N-morpholino)propanesulfonic acid with 50 mM sodium acetate and 10 mM EDTA adjusted to pH 7.0
260 μL HCHO
180 μL DEPC-$H_2O$
100 μL 80% glycerol
80 μL saturated bromphenol blue

The tube is stored at –70 °C.

Step 25. A tube containing 34 μL of mouse *Pal* cDNA (34 ng) is heated at 95 °C for 5 min, placed on ice immediately for 5 min, and centrifuged 5 sec to bring all solution into the bottom of the tube. To this tube are added

10 μL 5X NEB Phototope Labeling mix (buffers and primers)
5 μL dNTP mix (1mM each dNTP)
1 μL Klenow fragment of DNA Polymerase I (5 Units).

The random octamers and the dATP are biotinylated. The labeling reaction is incubated at 37 °C overnight.

80. What is the purpose of the Klenow fragment of DNA Polymerase I?

a. synthesis of labeled DNA from the cDNA template
b. the synthesis of additional first-strand DNA from RNA
c. filling in breaks in the *Pal* cDNA
d. rejoining the ends of the severed pBR322
e. inserting the *Pal* cDNA into the circularized pBR322

81. Why are the random octamers biotinylated?

a. That is the only substrate the Klenow fragment of DNA polymerase will accept.
b. Biotinylation keeps the two single strands of *Pal* cDNA from re-annealing.
c. The biotin groups will react with the streptavidin in a later detection reaction.
d. Biotinylation keeps the octamers from self-associating.
e. Biotinylation keeps adenine from being added on to the 3′ end of the octamer.

Step 26. After overnight incubation of the tube from step 25, above, 5 μL 0.2 M EDTA, pH 8.0, is added to stop the labeling reaction. Next, 5 μL 4 M LiCl and 100 μL 100% ice-cold ethanol are added. This solution is vortexed and stored at –70 °C for at least 30 min to precipitate the labeled DNA. The tube is centrifuged at 4 °C for 30 min at 16,000 × g to pellet the DNA. The ethanolic supernatant is then removed, and 500 μL ice-cold 70% ethanol is added. The tube is inverted several times to wash the pellet. The tube is centrifuged at 16,000 × g at 4 °C for 10 min. The ethanolic supernatant is removed and the pellet is allowed to air dry. The pellet is resuspended in 20 μL of 1X TE buffer (10 mM Tris-HCl, pH 8.0; 1 mM $Na_2EDTA$). The *Pal* biotinylated probe is stored at –20 °C.

82. True or false: Lithium chloride (LiCl) precipitates large DNA molecules, but not small oligonucleotides.

Step 27. To 100 mL melted agarose containing MOPS/EDTA, now add 5.4 mL 37% formaldehyde, mix, and pour two gels. Allow both gels to sit for at least 1 h.

Step 28. To the tube of "Total RNA" (step 8, above), add 14 μL of Northern Loading Dye and mix.

Step 29. Denature the sample of poly A+ RNA, the total RNA, and the mouse spleen poly A+ RNA in Northern Loading Dye by heating for 2 min at 65 °C and place on ice immediately for at least 1 min. To each of the 3 RNA tubes add 1 μL of 1 mg/mL ethidium bromide.

Step 30. Load the three RNA samples into the formaldehyde gel in the following order:

Spleen poly A+ RNA (16 μL in Northern Loading Dye, 0.5 μg)
Skip
Liver Total RNA (16 μL in Northern Loading Dye, 0.5 μg)
Skip
Liver poly A+ RNA (16 μL in Northern Loading Dye, 0.5 μg)

Electrophorese at 50 VDC at room temperature for 2 h.

83. True or false: Formaldehyde acts as a denaturing agent in the electrophoresis described above.

Step 31. Photograph the gel and mark where the wells are located. Save the gel.

Step 32. Cut a piece of nylon membrane (Immobilon – Millipore) to the exact same size as the gel. Pre-wet the membrane first with DEPC-$H_2O$ (minimum: 5 min) and then 10X SSC (1.5 M NaCl, 0.15 M sodium citrate, pH 7.0). Cut three pieces of Whatman 3MM filter paper to the same size as the gel and two pieces the same width as the gel but 3" longer to form a wick at the top and the bottom of the gel.

Step 33. Saturate the Whatman papers from step 32, above, with 10X SSC and place on a transfer support such that the ends of the wick dip into 10X SSC. Place the gel face down on the wick papers. Place the moistened nylon membrane on the gel. Place one pre-moistened piece of Whatman paper on the membrane. Add two dry pieces of Whatman paper to the stack. Place a stack of paper towels that is at least 3" thick on the filter papers. Place a rigid support on the stack of paper towels and a 200–500 g weight on the rigid support. Cover the apparatus with plastic wrap and allow the transfer to proceed overnight at room temperature.

84. What is the above transfer procedure known as?

a. a Northern blot
b. a Southern blot
c. an Eastern transfer
d. a Western blot
e. None of the above since this transfer has no common name.

Step 34. Disassemble the transfer apparatus and rinse the membrane in 6X SSC. Check the gel for any remaining fluorescence (this would indicate a problem with the transfer). Allow the membrane to air dry and cross-link with UV light. Next bake the membrane between two pieces of filter paper for 1 h at 80 °C.

Step 35. RNA-side out, place the membrane in a hybridization bag and add sequentially:

6 mL of Phototope Hybridization solution (4X SSPE, which is 3 M NaCl, 0.2 M $NaH_2PO_4$, 20 mM EDTA, pH 7.4)
1% sodium dodecyl sulfate or SDS
50X Denhardt's solution, which is 2.5 g Ficoll, 2.5 g polyvinylpyrrolidone, 2.5 g bovine serum albumin Fraction V in 250 mL of $H_2O$
1 mg/mL denatured, sheared salmon sperm DNA
50% formamide

Pre-hybridize at 57 °C with gentle rocking for 3–6 h.

85. What is the purpose of the formamide?

a. to increase the $T_m$ of any DNA/RNA complexes
b. to decrease the $T_m$ of any DNA/RNA complexes
c. to increase the density of the solution to make adding to a "submarine gel" well easier
d. to block "sticky spots" on the membrane
e. to solubilize the polyvinylpyrrolidone

Step 36. Retrieve the biotinylated probe from step 6, above. Add 30 µL of Phototope Hybridization solution and denature at 95 °C for 10 min. Place on ice immediately after denaturation. Centrifuge briefly to force all of the solution to the bottom of the tube.

Step 37. Retrieve the pre-hybridization bag from step 35, above, cut the bag, and pour out the pre-hybridization solution. Pipette 6 mL of Phototope Hybridization solution (4x SSPE, 1% SDS, 5X Denhardt's solution, 5% dextran sulfate, and 50% formamide) into the hybridization bag. Next add all 50 μL of denatured probe to the hybridization bag and re-seal the bag. Hybridize overnight at 57 °C with gentle rocking.

Step 38. Cut open the hybridization bag from step 37, above, and wash the membrane in 500 mL of 2X SSPE/0.1% SDS. Repeat the wash with shaking at 57 °C for 5 min. Repeat the 57 °C wash. Wash for another 5 min at 57 °C with rocking and using 0.1X SSPE/0.1% SDS. Repeat the 57 °C, 0.1X SSPE/0.1% SDS wash and then give the membrane a final room temperature wash with 0.1X SSPE/0.1% SDS.

Step 39. Place the membrane from step 38, above, in a new plastic bag with a spout and add 6 mL of Phototope Blocking solution. Seal the spout and rock for at least 5 min at room temperature. Add 4 mL of streptavidin solution to the bag, re-seal, and rock for an additional 5 min at room temperature. Add 30 mL of Phototope Wash solution 1 and incubate for 5 min with rocking at room temperature. Discard the wash and repeat with a fresh 30 mL of Phototope Wash solution. Add 4 mL of alkaline phosphatase solution to the bag and incubate for 5 min with rocking. Discard the alkaline phosphatase solution and replace with 30 mL of Phototope Wash Solution 2. Incubate for 5 min with rocking at room temperature. Repeat the final wash and leave it in the bag until ready for the next step.

86. True or false: The main detection reaction in the above set of steps is the well-known biotin-avidin reaction.

Step 40. Discard the last wash and add 4 mL of Phototope Detection Mix (3.84 mL $H_2O$ + 160 μL 25X Star Diluent + 40 μL Detection Reagent). Incubate the Detection Mix with rocking at room temperature for 5 min.

Step 41. Wash the outside of the bag with tap water and dry. Expose to film for 10–40 min. At this point, a Northern Blot lumigram will have been obtained and can be examined. Also, at this point, an alternative to the Northern Blot, an RT-PCR, will be described in the following steps.

Step 42. Using another 1 μL of mouse liver sample, set up two reverse transcriptase reactions for a quantitative RT-PCR experiment—one using 100 ng of total RNA isolated from mouse liver and one using 100 ng of mouse spleen. The 3X reverse transcriptase mixture (18 μL for each tube with the remainder being discarded) is described below:

16.5 μL DEPC-treated $H_2O$
13.5 μL $MgCl_2$
12 μL dNTP mix
6 μL 10X RT buffer
3 μL random primers
1.5 μL RNase inhibitor
1.5 μL MultiScribe RT (50 U)

The RT PCR part of the reaction itself is given in the following steps:

25 °C for 10 min
48 °C for 30 min
95 °C for 5 min

87. True or false: The third step in the above series is required to deactivate the reverse transcriptase.

Step 43. At the end of the reaction, place the mixture on ice.

Step 44. Next, set up the PCR for each tube using the following:

75 μL 2X Taq Man Universal Master Mix (also contans uracil-N-glycosylase or UNG)
15 μL *Pal* Probe (Fam dye), 2 μM
12 μL $H_2O$
9 μL *Pal* Forward T primer (10 pmol/μL)
9 μL *Pal* Reverse T primer (10 pmol/μL)

88. True or false: The universal Master Mix in the above reaction contains dTTP, not dUTP.

The PCR is set for the following:

50 °C for 2 min
95 °C for 10 min
40 cycles of 95 °C for 15 sec and 60 °C for 1 min

The setup also includes a no-template, negative control.

89. What is *true* about the first step in the above PCR?

a. It is optimal for annealing probe to DNA.
b. It is optimal for denaturing probe from DNA.
c. It is optimal for second-strand synthesis.
d. It is optimal for UNG activity.
e. It activates the fluorescence of the Fam dye.

90. What might exposure of DNA to UV light produce?

a. thymine-thymine cyclobutane dimers between adjacent thymines
b. [6,4] products between adjacent thymine-cytosine pairs
c. [1,6] products between adjacent guanosine–adenine pairs
d. all of the above
e. only *a* and *b*

91. True or false: It is often implied that genes are dominant or recessive. This is incorrect. It is disease states or phenotypes that are dominant or recessive.

92. In the time-course of an HIV infection, when does the HIV viral load tend to be the highest?

a. on initial infection or at exposure
b. about 2–4 weeks after exposure
c. about 3–4 months post-exposure
d. about 8–10 years post-exposure
e. Either *b* or *d* may be correct.

93. Which of the following is/are *true* about the major predisposing gene in Peutz-Jeghers syndrome (PJS)?

a. It is *LKB1*.
b. It codes for a serine/threonine kinase.
c. It is found as a mutation in most, but not all, PJS patients.
d. All of the above are true.
e. Only *b* and *c* are true.

94. How does the DEB (diepoxybutane) test used to diagnose Fanconi's anemia use DEB?

a. to form [6,4] products between adjacent thymine-cytosine pairs
b. as a mitogen
c. to stimulate unscheduled DNA synthesis
d. to measure the cell's sensitivity to chromosomal breakage by crosslinking agents
e. to induce sister chromatid exchange

95. Consider the segment of dsDNA

5′-$(N'_x)_{1500}$-A T G C C T T A A T G C C T A G A G A A T C C G T A G –$(N'_y)_{2300}$-3′

3′-$(N_x)_{1500}$-T A C G G A A T T A C G G A T C T C T T A G G C A T C-$(N_y)_{2300}$-5′

where $N_x$ and $N'_x$ and $N_y$ and $N'_y$ are any two complementary pairs of sequences of nucleic acids. PCR amplification of the sense sequence ATGCCTTAATGCC-TAGAGAATCCGTAG and its anti-sense sequence shown above is required using the forward primer 5′-ATGCCTTAAT-3′ and the reverse primer 5′-CTAC-GGATTC-3′. Assuming the correct reaction conditions, one starting template dsDNA, and a 100% theoretical yield, after 20 cycles which of the following would be obtained?

a. only 524,288 copies of the desired amplicon
b. the original segment of DNA plus 524,288 copies of the amplicon
c. the original segment of DNA, plus 20 single strands of 5′-A T G C C T T A A T G C C T A G A G A A T C C G T A G –$(N'_y)_{2300}$-3′, 20 single strands of 3′-$(N_x)_{1500}$-T A C G G A A T T A C G G A T C T C T T A G G C A T C-5′, and 1,048,576 copies of the desired amplicon
d. only 1,048,576 copies of the desired amplicon
e. the original segment of DNA, plus one single strand of 5′-A T G C C T T A A T G C C T A G A G A A T C C G T A G –$(N_y)_{2300}$-3′, one single strand of 3′-$(N_x)_{1500}$-T A C G G A A T T A C G G A T C T C T T A G G C A T C-5′, and 524,288 copies of the desired amplicon

**96–99. Roche Amplicor Assay for *M. tuberculosis***

96. In the Roche Amplicor assay for *M. tuberculosis*, what is the target for amplification after specimen lysis?

a. MTB ssDNA
b. 16S MTB rRNA
c. MTB dsDNA
d. MTB mRNA
e. MTB tRNA coding for the initial methionine

97. Why is AmpErase added to the “master mix”?

a. to remove any uracil-containing amplicon (synthetic amplicon) that may be present in amplicon from previous assays or carryover
b. to terminate extraneous extensions with a dUTP
c. to hydrolyze nascent mRNA
d. to hydrolyze nascent rRNA
e. to terminate extraneous extensions with dATP

98. How much laboratory space does the Roche Amplicor assay require?

a. only a single room
b. two separate rooms
c. three separate rooms
d. Either *b* or *c* may be true, depending on the assay.
e. Either *a* or *b* may be true, depending on the assay conditions.

99. True or false: An internal control may be added to the Roche Amplicor assay to determine whether an inhibitor is or is not present.

100. In the technique known as ASO (allele specific oligonucleotide) hybridization, where is the label?

a. in the PCR reaction spotted onto the membrane
b. ini the PCR reaction used to probe the membrane
c. in the oligonucleotide spotted onto the membrane
d. in the oligonucleotide used to probe the membrane

101. In the technique known as *Reverse* ASO (allele specific oligonucleotide) hybridization, where is the label?

a. in the PCR reaction spotted onto the membrane
b. in the PCR reaction used to probe the membrane
c. in the oligonucleotide spotted onto the membrane
d. in the oligonucleotide used to probe the membrane

102. Why must special care be taken when working with acrylamide gel solutions?

a. to prevent the inadvertent contamination of the gel solution with foreign DNA
b. because acrylamide is a potent neurotoxin
c. because acrylamide can generate free radicals that can start a fire
d. to prevent premature discoloration of the solution, which might obscure results

103. Polyacrylamide DNA gels are often run under denaturing conditions to keep the strands separated. Which of the following agents is frequently used to provide this denaturing condition?

a. 8 M urea
b. 100 mM NaCl
c. 10 mM Tris, 0.2 mM EDTA, pH 8.3
d. bis-acrylamide

104. In SSCP analysis, is the polyacrylamide gel run under native or denaturing conditions?

a. native
b. denaturing

105. DNA is PCR-amplified for both a known normal individual and a patient with a suspected mutation. Both PCR products are mixed together, denatured, and allowed to reanneal. The reannealed products are then separated on a specially formulated high-resolution polyacrylamide gel. Which procedure is this?

a. denaturing gradient gel electrophoresis (DGGE)
b. single-stranded conformational polymorphism analysis (SSCP)
c. heteroduplex analysis (HA)
d. restriction fragment length polymorphism analysis (RFLP)

106. What is the principle behind DGGE?

a. Heteroduplexes migrate slower than do homoduplexes.
b. DNA melting occurs in domains that are less stable; molecules with melted regions migrate more slowly than molecules that are completely intact.
c. Mutations cause changes in restriction digestion patterns.
d. Several PCR reactions can be run simultaneously in a single tube.

107. Forward Primer A is a sequence-specific primer for a wild-type gene, and Forward Primer B is a sequence-specific primer for a mutated gene. Predict the effect on the PCR efficiency if Primer B were used to amplify a wild-type gene sequence (NOTE: The reverse primer is not shown).

**Primer A: GTGCGTGACATTAAGGAGAAG**

**Primer B: GTGCGTGACATTAAGGAGAAC**

a. twice normal efficiency
b. normal efficiency
c. one-half normal efficiency
d. No PCR product will be formed.

108. This gel image is of a multiplex PCR assay to detect deletions in the dystrophin gene; nine separate regions spanning the entire gene were amplified using exon-specific primers. Which individual is most likely ***not*** to be affected with the disorder?

a. D
b. E
c. F
d. G

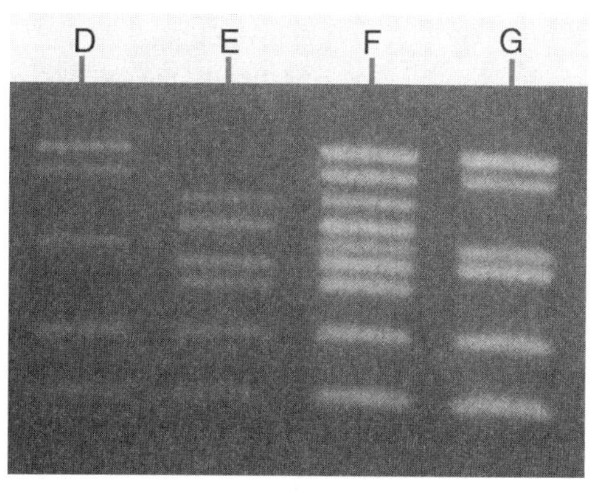

109. Refer to question 108 and its accompanying gel image. Why are the first, second, and fifth bands in Lane F missing from lane E?

a. The PCR amplification failed because of primer misannealing.
b. These exons have been deleted in this individual.
c. The reverse transcriptase was inactivated due to excessive heating.
d. Insufficient dNTPs were added to the reaction.

110. What modifications to a standard PCR are *not necessary* in order to perform a multiplex PCR?

a. increased concentration of dNTPs
b. increased concentration of *Taq* polymerase
c. addition of exonuclease III
d. adjust the extension time at 72 °C

111. Why is the Southern blot technique called the Southern blot technique?

a. It was developed in South Africa.
b. It is the opposite of the Northern blot technique, developed by E.M. Northern in 1975.
c. It works best at higher temperatures (such as those commonly found in the South).
d. It is named after E.M. Southern, who developed the technique in 1975.

112. What is the purpose of the Southern blot?

a. to transfer RNA from an electrophoresis gel to a membrane
b. to transfer DNA from an electrophoresis gel to a membrane
c. to transfer protein from an electrophoresis gel to a membrane
d. to hybridize radioactive probe to bound nucleic acids

113. Visualization of electrophoresed DNA in an agarose gel is often accomplished through the use of the fluorescent dye molecule, ethidium bromide (EtBr). How is this accomplished?

a. EtBr intercalates between the strands of dsDNA and becomes highly fluorescent when exposed to UV light.
b. EtBr stains the dsDNA a bright blue color, visible during electrophoresis.
c. EtBr covalently attaches to dsDNA and becomes fluorescent when exposed to visible light.
d. EtBr melts the dsDNA strands; the resultant ssDNA is fluorescent when exposed to UV light.

114. Which of the following statements is *true*?

a. EtBr binds more strongly to ssDNA than to dsDNA.
b. EtBr binds more strongly to dsDNA than to ssDNA.
c. EtBr does not bind at all to ssDNA.
d. EtBr does not bind at all to dsDNA.

115. Which of the following methods of DNA transfer from an agarose gel to a nylon membrane is the *cheapest*?

a. capillary transfer
b. vacuum transfer
c. electroblotting
d. none of the above

116. You are called upon to analyze a limited sample of DNA for the presence of a very rare gene. You have electrophoresed your sample through an agarose gel and are getting ready to transfer the DNA to a membrane. Which transfer method would you choose, and why?

______________________________________________

______________________________________________

______________________________________________

117. What is DNA that is best able to be transferred to a membrane best described as?

a. single stranded
b. double stranded
c. triple stranded (H-DNA)
d. UV crosslinked

118. The complete DNA sequence of the bacterium *H. influenzae* was determined by which sequencing assembly method?

a. the shotgun approach
b. the clone contig approach
c. the chromosome-by-chromosome approach
d. the Sanger approach

119. In the vacuum transfer blotting technique, if a vacuum of 20 in-Hg transfers DNA in 3 h, would you expect that increasing the vacuum strength to 60 in-Hg would shorten the time to 1 h?

a. Yes, increasing vacuum strength speeds up the transfer rate and efficiency.
b. No, increased vacuum strength can collapse the gel and reduce the transfer rate and efficiency.

120. Following transfer of DNA from a gel to a membrane, what is the next step?

a. development of the colorimetric detection reaction
b. hybridization with a labeled probe
c. hybridization with an unlabeled probe
d. denaturation of the DNA

121. What is the recommended buffer for the electrophoresis of RNA?

a. 1X TBE, pH 8.3
b. 1X MOPS, pH 7.0
c. 1X HEPES, pH 8.9
d. 0.2 M NaOH, 50 mM NaCl

122. Which of the following agents may be used to provide denaturing conditions during RNA electrophoresis?

a. formaldehyde
b. glyoxal
c. 0.2 M NaOH, 50 mM NaCl
d. both *a* and *b*

123. When transferring RNA from a gel to a membrane, which transfer buffer is best?

a. 10X TBE, pH 8.3
b. 20X SSC (salt-sodium citrate), pH 7.0
c. 0.5 M NaOH
d. none of the above

124. You have set up a capillary transfer from an RNA gel to a charged nylon membrane; however, you have inadvertently left a small bubble between the gel and the membrane. What sort of effect will this have on the experiment?

a. None. Bubbles will simply rise to the top of the paper towel stack.
b. None. RNA will transfer through the bubble slower than usual, but overnight incubation will overcome this difficulty.
c. Large effect. RNA can not transfer through a bubble.
d. Large effect. Bubbles will degrade RNA.

125. What is the *Klenow fragment*?

a. a modified *E. coli* DNA polymerase lacking the 5′→3′ exonuclease activity
b. a highly repetitive stretch of DNA located near the centromeres of chromosomes
c. the thermostable enzyme used in PCR
d. the common name for AMV reverse transcriptase

126. In the *Nick Translation* reaction, how is label incorporated into the probe?

a. through the exchange reaction where the 5′-phosphate at the nick is exchanged for the $\gamma$-phosphate of ATP
b. via incorporation of labeled dNTPs
c. through extension of a labeled primer
d. through the action of alkaline phosphatase

127. A double-stranded DNA template is denatured and mixed with random sequence octamers at room temperature. The primers are then extended by the *Klenow fragment*. Which procedure is being described here?

a. the production of riboprobes
b. nick translation
c. random primer labeling
d. end labeling

128. 5′ End-labeling of oligonucleotide primers is accomplished by the use of which enzyme?

a. T4 DNA ligase
b. T4 polynucleotide kinase
c. T4 endonuclease V
d. T7 RNA polymerase

129. You wish to probe a Northern blot and want to use a probe with the strongest possible binding. What type of probe would you most likely use?

a. a nick-translated DNA probe
b. a random primed DNA probe
c. a strand-specific riboprobe
d. an end-labeled oligonucleotide

130. You have probed a Southern blot with a strand-specific riboprobe labeled with $^{32}P$ and wish to rehybridize the membrane with a new probe. What do you do?

a. Prepare a new probe and rehybridize immediately; the old probe will not interfere.
b. Remove the first probe by treating the membrane with NaOH and heat.
c. Wait six months for the radioactivity in the first probe to decay and then rehybridize with the new probe.
d. Rehybridize immediately with a new probe using a different type of label, such as fluorescent or chemiluminescent.

131. In a FISH procedure, which of the following types of probes will specifically light up the centromere of the chromosome?

a. unique sequence probe
b. chromosome paint
c. satellite probe
d. none of the above

132. 30 cycles of PCR of genomic DNA in the test tube can yield $2^{30}$ copies of amplified target. The same amplification performed *in situ* (that is, to fixed tissue on a slide) may be expected to yield how many copies of amplified product?

a. $2^{30}$; The reaction is just as efficient.
b. 50–200; The reaction is much less efficient.
c. $2^{40}$; The reaction is much more efficient.
d. 0; PCR will not work *in situ*.

133. What are the short repetitive sequences that occur at the telomeres and centromeres of human chromosomes called?

a. satellite sequences
b. minisatellite sequences
c. microsatellite sequences
d. polygenic sequences

134. Which sequences are also known as VNTRs?

a. satellite sequences
b. minisatellite sequences
c. microsatellite sequences
d. polygenic sequences

135. Which sequences are responsible for such genetic illnesses as Fragile X syndrome and Huntington disease?

a. satellite sequences
b. minisatellite sequences
c. microsatellite sequences
d. polygenic sequences

**136–171.** The goal of the following experiment is to construct a size-selected mouse genomic DNA library from which clones of a specific, highly-repeated mouse DNA sequence (part of the LINE-1 repeat) can be isolated. A condensation of the steps in the experiment is described below:

Step 1. A 40-μL (5 μg) aliquot of mouse genomic DNA in TNE buffer is mixed with 5 μL of 10X EcoRI buffer; 4 μL $ddH_2O$ (sterilized, double-distilled water); and 1 μL EcoRI enzyme that has been stored at –20 °C. The mixture is stirred with the pipette tip, centrifuged for 5 sec in a nanofuge, and digested overnight at 37 °C. After overnight incubation, a second 1 μL aliquot of EcoRI enzyme is added and the mixture is incubated for 3 h at 37 °C. After incubation, the digest is extracted with organic solvents and the volume reduced to 20 μL using a speed-vacuum concentrator or ethanol precipitation.

136. Cutting the mouse DNA with EcoRI leaves what?

a. blunt ends with a G at the 5′ end and a C at the 3′ end
b. 5′-AATT overhangs on both ends
c. 5′-ATTC overhangs on both ends
d. 5′-TAA overhangs on both ends
e. TTAA on the 5′ end and AATT on the 3′ end

137. Why must additional EcoRI enzyme be added after the overnight incubation?

a. The enzyme denatures during the overnight digestion, reducing its activity and requiring that additional enzyme be added to complete the digestion.
b. The digestion process creates 100 kb fragments that bind irreversibly to the EcoRI active site, deactivating the enzyme.
c. The digestion process frees mouse ribozymes from the dsDNA, and the ribozymes proteolyze the EcoRI enzyme.
d. The digestion process frees mouse ribozymes from the ssDNA, and the ribozymes proteolyze the EcoRI enzyme.
e. Nucleotides that are freed during the digestion bind irreversibly to the EcoRI catalytic site, deactivating the enzyme.

Step 2. An agarose gel is prepared in a microwave oven from 1.0 g low-melt agarose and 100 mL of 1X Tris-acetate buffer (TAE). The gel is cast using a comb that will provide the wells and covered with 1X TAE running buffer.

138. True or false: The low-melt agarose gel is sturdier and easier to handle than regular agarose gels.

Step 3. The concentrated digest from Step 1, above, is heated to 65 °C for 20 min and mixed with 5 μL of "Blue Juice" (20% ficoll; 0.01 M Tris-HCl, pH 8.0; 1 mmol $Na_2$-EDTA; and 0.1% xylene cyanol FF 0.1% bromphenol blue).

139. True or false: The purpose of the ficoll is to make the "Blue Juice" heavier than a buffer solution and cause sample added to a gel to "sink" into the well.

140. Why is the concentrated digest heated to 65 °C?

a. to re-anneal single strands of DNA into double-stranded DNA
b. to melt dsDNA into ssDNA
c. to inactivate the EcoRI enzyme
d. to evaporate any excess single nucleotides produced in the digestion process
e. to evaporate any free buffer base

Step 4. The low-melt agarose gel is loaded using the following schedule:

Lane 3: 6 μL (150 ng) of λ DNA that has been cut with HindIII (pre-mixed with "Blue Juice" and pre-heated to 65 °C and cooled on ice).
Lane 4: 20 μL (200 ng) uncut mouse DNA in "Blue Juice."
Lane 6: 25 μL EcoRI digested mouse DNA in "Blue Juice."
Lane 7: 6 μL (150 ng) of λ DNA that has been cut with BstEII (pre-mixed with "Blue Juice" and pre-heated to 65 °C and cooled on ice).

The next step is to electrophorese ("run out") the gel for 2 h at 60 VDC, remove the gel from the electrophoresis chamber, and stain the gel for 8 min with ethidium bromide (300 μL of a stock 10 mg/mL solution in $ddH_2O$ per 500 mL of water).

141. What is the purpose behind heating the Hind III- and BstE II-cut DNA?

a. to separate the cohesive ends of fragments 1 and 4
b. to cause all ssDNA to form dsDNA
c. to separate the cohesive ends of fragments 2 and 3
d. to melt all dsDNA into ssDNA
e. to separate the cohesive ends of fragments 5 and 8

Step 5. Under long-wavelength UV light, carefully cut out the 1.4 kb EcoRI mouse repeat band from the gel, place the cut-out section of the gel into a microfuge tube, and discard the remainder of the gel. Briefly centrifuge the microfuge tube to bring all of the gel material to the bottom of the tube (NOTE: The 1.4 kb EcoRI mouse repeat is a subsection of the 7 kb LINE-1 repeat that contains the mouse reverse transcriptase gene (Rvt). This repeat exists in several hundred thousand copies in the mouse genome and contains two EcoRI sites that are 1.4 kb apart. Because of the high copy number of the LINE-1 repeat, the 1.4 kb band is visible in the digested smear of the mouse genomic DNA).

142. What is the purpose behind using long-wavelength UV light?

a. to minimize damage to the DNA
b. to keep the DNA single stranded
c. to keep the DNA double stranded
d. to prevent re-annealing of the overhang ends of the cut DNA
e. simply because long-wavelength UV lamps are cheaper than short wavelength UV lamps

Step 6. Melt the agarose in the microfuge tube from Step 5, above, at 70 °C for 10 min and raise the total volume in the tube to approximately 120–170 μL with $ddH_2O$. Return the tube to 70 °C for 5 min.

143. What is the purpose behind using 70 °C?

a. to keep the DNA in a double-stranded conformation
b. to keep the DNA in a single-stranded conformation
c. to prevent re-annealing of the overhang ends of the cut DNA
d. to denature any remaining restriction endonuclease
e. to melt the agarose without causing the double-stranded DNA to melt or disassociate into single strands

Step 7. Lower the temperature of the microfuge tube to 41 °C, equilibrate for 5 min, add 20 μL of 10X beta-agarase digestion buffer and 10 μL of beta-agarase enzyme, and incubate at 41 °C for 60 min.

144. What is the purpose of lowering the temperature to 41 °C?

a. to allow the agarose to partially re-gel
b. to allow the beta-agarase to digest the agarose without allowing the agarose to re-gel
c. to allow the single strands of DNA to re-nature into double-stranded DNA
d. to allow the single strands of DNA to adsorb to the agarose, which protects the DNA from denaturing
e. to allow methylation of cytosine residues in the DNA through a transferase activity

Step 8. Place a Microcon-100 (molecular weight cut-off 125 bp for dsDNA and 300 nucleotides for ssDNA) sample reservoir into a Microcon vial, load the reservoir with 200 μL of the beta-agarase digest from Step 7, above, and 300 μL of 15% isopropanol, and centrifuge for 14 min at 500 × g. Discard the eluant. Add 450 μL of $ddH_2O$ to the reservoir and centrifuge for 14 min at 500 × g. Discard the column eluant and repeat. Recover the DNA by inverting the filter and centrifuging for 2 min at 500 × g. Reduce the DNA solution to complete dryness using a speed-vac concentrator.

145. Which is *true* about the Microcon procedure?

a. It allows agarose fragments (and other low-molecular-weight impurities) from the digestion to pass through while retaining the DNA, which is collected after the filter is reversed.
b. It retains agarose fragments and DNA, both of which are collected after the filter is reversed.
c. It retains dsDNA and allows all ssDNA to pass through.
d. It retains ssDNA and allows dsDNA to pass through.
e. It does not recover the DNA. The procedure actually stops here.

Step 9. Re-suspend the 1.4 kb mouse DNA from step 8, above, in 7 μL of $ddH_2O$ and add all of it to 1 μL of EcoRI-cut and dephosphorylated λZAP Express vector DNA (1.0 μg). To this mixture add 1 μL of 10X ligase buffer and 1 μL T4 DNA ligase. Incubate the ligation mixture overnight at 14 °C.

146. True or false: The λZAP Express vector DNA has 1 EcoRI site just like the wild type.

Step 10. Package the mouse-λZAP Express ligation reaction products by immediately adding 5 μL of the reaction from Step 9, above, to lambda packaging extract that has just thawed. Mix by stirring with a pipette tip. Centrifuge the mixture for 5 sec to make certain that all reactants are at the bottom of the tube and incubate for 2 h at room temperature. Following the 2-h incubation, add 500 μL of SM buffer (phage storage buffer) and 20 μL of $CHCl_3$ to what is now the "Genomic Library" tube.

147. True or false: The packaging process places the ligated DNA into functional bacteriophage lambda that consists of the bacteriophage head and tail fibers. The bacteriophage self-assembles upon thawing and with the addition of the ligated DNA.

Step 11. Dilute a portion of the "Genomic Library" 1:10 by removing 1 μL to a fresh tube and adding 9 μL of SM buffer. Add 2 μL of the dilute "Genomic Library" to 200 μL of XL1-Blue MRF *E. coli* cells. To a separate tube containing 200 μL of XL1-Blue MRF *E. coli* cells, add 2 μL of the undiluted "Genomic Library." Mix and incubate both tubes for 15 min at 37 °C.

Step 12. To two tubes containing 3 mL of NZY top agarose in a tube, add 15 μL of 0.5 M IPTG (iso-propyl-β-D-thiogalactopyranoside) at 55 °C. To each tube add 50 μL of 250 mg/mL X-gal (5-bromo-4-chloro-3-indoyl-β-D-galactoside). Add either the undiluted or diluted suspension from Step 11, above, and immediately pour into a petri dish pre-coated with agar medium, allow the top agarose to solidify, and place in a 37 °C incubator inverted for 8–9 h.

148. True or false: The bacteriophage lambda DNA that is injected into the *E. coli* cells is linear but quickly circularizes to prevent degradation.

149. True or false: Using a gross excess of *E. coli* cells to phage (100–1000X) helps prevent double infection of cells and insures that a smooth "lawn" of *E. coli* cells is formed.

Step 13. Count the number of total plaques (clear spots) on each plate. The titer is the total number of plaques divided by 2, multipled by 10 if the diluted plate is counted, and multipled by the total volume of the library (530 μL). Also, determine the total number of clear (recombinant) plaques. This yields the total number of plaque-forming units (pfu) in the library.

150. A diluted sample plate has 50 plaques. Calculate the number of pfu's (plaque-forming units) in the library. ________________________________

151. Wherever there is a clear spot or plaque,

a. phage failed to infect *E. coli.*
b. phage infected *E. coli*, but cell lysis did not occur.
c. phage infected *E. coli* and cell lysis did occur.
d. *E. coli* successfully lysed phage.
e. *E. coli* infected phage, but did not lyse it.

Step 14. Either repeat the above plating procedure with a sample diluted such that about 50 plaques per plate will be obtained, or use a plate from Step 13, above, that has approximately 50 plaques for a plaque lift, by first refrigerating the plate. Label an S&S Nytran+ nylon filter with a pencil and place three unique markers asymmetrically on the label side of the filter. Taking care to avoid air bubbles, place the nylon membrane writing side down on the top agarose. Copy the unique, asymmetric marks onto the bottom of the pertri dish. Without disturbing the agarose, remove the nylon filter and place plaque-side up on a piece of filter paper to dry. Denature the filter for 2 min on filter paper saturated with 0.5 N NaOH and 1.5 M NaCl, then neutralize on paper saturated with 0.5 M Tris-HCl/pH 7.5, 1.5 M NaCl. Place this on paper saturated with 2X SSC for 2 min and air dry for 15 min. UV crosslink and bake at 80 °C, and place the plaque-lift filters into plaque wash buffer (50 mM Tris, pH 8.0; 1 M NaCl; 10 mM EDTA; and 0.1% SDS) for 2–4 h at 42 °C.

152. What is the purpose of the NaOH?

a. to lyse phage and convert ssDNA into dsDNA
b. to lyse remaining *E. coli* and convert ssDNA into dsDNA
c. to lyse phage and convert dsDNA into ssDNA
d. to lyse remaining *E. coli* and convert dsDNA into ssDNA
e. to completely break all of the amide bonds in the nylon support in order to provide binding sites for the DNA

153. What is the purpose of the UV crosslinking and baking step?

a. to form nucleotide dimers
b. to convert ssDNA into dsDNA
c. to break down the amide bonds in the nylon filter for later dissolution to free the fixed DNA
d. to fix the DNA to the nylon filter
e. to form irreversible chemical bonds between the nucleotides that constitute the dsDNA

154. True or false: The plaque lift can be performed only once per plate.

155. True or false: The drying step should be performed in a vacuum oven because nylon, which is an organic polymer, will degrade at 80 °C just like nitrocellulose.

Step 15. Pre-hybridize the plaque lift filter with Probe-Amp hybridization buffer (5X SSC, 5.0% dextran sulfate, 0.5% nonfat dry milk, 0.1% SDS, and 100 μg/mL denatured/sheared salmon sperm DNA) for 1–6 h at 65 °C.

156. What is the purpose of the salmon sperm DNA?

a. to block “sticky” spots on the membrane so that the probe used in the next step will anneal only to the DNA on the membrane and not stick to the random spots on the membrane
b. to bind to the DNA that is linked to the plate so that only a high-affinity probe can displace the DNA
c. to form histone-like complexes with the protein in the nonfat dry milk
d. to provide the correct environment for the binding probe in the next step
e. to be broken down into single nucleotides to be used in the PCR step

Step 16. Prepare a labeled (non-radioactive) mouse 1.4 kb Rvt (Eco) repeat DNA probe from the following PCR reaction containing the plasmid pBluescript with one cloned copy of the mouse 1.4 kb Rvt (Eco) repeat (4 ng in 29.5 μL):

8 μL 25 mM $MgCl_2$
5 μL 10X PCR buffer
5 μL FL-dUTP PCR mix where FL is fluorescein
1 μL RevT (Eco) repeat forward primer (10 pmoles/μL)
1 μL RevT (Eco) repeat reverse primer (10 pmoles/ μL)
0.5 μL *Taq* Gold DNA polymerase

The PCR program is a 30-cycle program the starts at 94 °C for 12 min and has 30 cycles of 94 °C for 60 sec, 50 °C for 60 sec, 72 °C for 90 sec, with a final 72 °C for 7 min and a hold at 4 °C.

157. Which is *true* about the initial "94 °C for 12 min" step?

a. It allows all components to dissolve completely.
b. It denatures the plasmid DNA.
c. It activates the *Taq* polymerase.
d. all of the above
e. only *b* and *c*

158. True or false: The final step (72 °C for 7 min) is an extension step.

Step 17. Mix 15 μL of the labeled DNA probe from Step 16, above, with 45 μL of hybridization buffer. Heat the probe for 5 min at 95 °C and place on ice. Add all 60 μL of the labeled probe to the pre-hybridization mixture from Step 15, above. Hybridize overnight at 65 °C with rocking. Remove the filter and wash in 1X SSC, 0.1% SDS, at 65 °C for 15 min with rocking, then 0.5X SSC, 0.1% SDS, at 65 °C for 20 min, and place in ECL buffer 1. Rinse the filter in 50 mL of Blocking Solution (ECL Buffer 1 with 0.5% nonfat dry milk) for 1 h or longer with rocking. Pour off the blocking solution and rinse briefly in ECL Buffer 1. Incubate the filter in 50 mL of antibody solution (1:1000 of anti-FL-HRP antibody that is anti-fluorescein covalently bonded to horse radish peroxidase) in 0.5% BSA in ECL buffer 1 for 1 h with agitation. Pour off the antibody solution and wash the filter with 0.1% Tween in ECL buffer 1 twice for 10 min each followed by two additional 5-min washes. Leave the filter in the final wash.

159. What is the purpose of heating the probe at 95 °C and then placing it on ice?

a. to activate the horse radish peroxidase (HRP)
b. to deactivate the HRP
c. to thermally place the fluorescein in an excited state and leave it there so that it will chemiluminesce later
d. to break the anti-fluorescein–fluorescein bonds so that the fluorescein can fluoresce later in the reaction scheme
e. to cause the double-stranded DNA probe to denature

Step 18. The wash covering the filter from Step 17, above, is poured off and replaced with a solution made from equal volumes of ECL detection solutions 1 and 2 in a darkroom. Incubate for 1 min. Drain the detection solution and place the filter into a development folder with the plaque side facing a sheet of x-ray film. Expose the film for 20–60 min, develop the film ("lumigram"), and look for positive plaques produced by the chemiluminescence (see lumigram below).

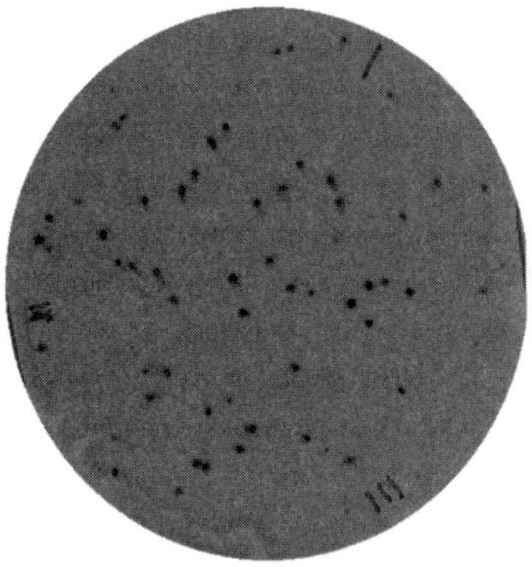

Step 19. Use the lumigram aligned with the petri plate from which it was derived to locate two single, isolated plaques that demonstrate a strong signal. Pick each plaque with a pipette tip and expel into 50 μL of SM buffer. Vortex the mixture and incubate for several hours at room temperature to release the phage from the agarose. Perform PCR on the mixture using 46 μL of the following Master Mix and conditions described below:

126 μL $ddH_2O$
32 μL dNTP mix
20 μL 10X PCR reaction buffer
12 μL 25 mM $MgCl_2$
4.0 μL T3 Primer (10 pmoles/μL)
4.0 μL T7 Primer (10 pmoles/μL)
2.0 μL *Taq* Gold Polymerase

For the PCR, an initial step of 94 °C for 12 min, 55 °C for 5 min and 35 cycles of 72 °C for 90 sec, 94 °C for 45 sec, and 55 °C for 45 sec. There is a final stage of 72 °C for 10 min, and then the PCR machine holds the samples at 4 °C.

160. In addition to activating the Taq Gold polymerase, what does the initial 94 °C step do?

a. ensures that the lambda phage are lysed and that the dsDNA is converted to ssDNA
b. also converts any circular DNA into linear ssDNA
c. also converts any circular DNA into linear dsDNA
d. disassociates any lambda phage DNA–RNA complexes
e. destroys any RNA

161. True or false: It is desirable for the $T_m$ for the $T_3$ and $T_7$ primers to be within at least 2 °C of each other.

Step 20. Transfer the PCR reaction mixture from Step 19, above, to a sturdier tube and add 250 μL of PB buffer. Mix by pipetting up and down and add to a QIAquick spin column with a 2 mL collection tube. Centrifuge the column and reservoir for 1 min at 10,000 × g. Discard the eluant and wash with 750 μL of Buffer PE. Discard the column wash. Centrifuge again to remove all of the Buffer PE and then elute the DNA with 30 μL of 0.1X TE, pH 8.0 by centrifugation at 10,000 × g for 1 min after allowing the buffer to sit on the column for 1 min.

162. Which is *true* in the clean-up step, above?

a. In the presence of high salt and pH ≤ 7.5, DNA, oligos, and dNTPs all bind to the silica to be eluted later while salts are eluted in the initial step.
b. In the presence of high salt and pH ≤ 7.5, DNA and oligos both bind to the silica to be eluted later while salts and dNTPs are eluted in the initial step.
c. In the presence of high salt and pH ≤ 7.5, DNA binds to the silica while oligos and dNTP's are eluted in the initial step.
d. In the presence of high salt and pH ≤ 7.5, oligos and dNTPs bind to the silica in the initial step while DNA is washed through the column in the initial centrifugation and wash.
e. In the presence of high salt and pH ≤ 7.5, dNTPs bind to the silica in the initial step while DNA and oligos are washed through the column in the initial centrifugation and wash.

163. True or false: Lowering the pH of the 0.1X TE will prevent elution of the DNA in the final step.

Step 21. Cast a 1% agarose gel. For the PCR product from Step 20, above, remove 5 μL and mix with 2 μL 5X "Blue Juice," and 3 μL $ddH_2O$. Similarly, mix 5 μL of the PCR product diluted 1:20 with $ddH_2O$ with 2 μL 5X "Blue Juice" and 3 μL $ddH_2O$. Load the agarose gel in the following sequence:

Lane 1: 10 μL undiluted PCR mixture.
Lane 2: 10 μL undiluted PCR mixture.
Lane 3: 10 μL undiluted PCR mixture.
Lane 4: 10 μL undiluted PCR mixture.
Lane 5: 10 μL diluted PCR mixture.
Lane 6: 10 μL diluted PCR mixture.
Lane 7: 6 μL λBstEII in "Blue Juice" after heating to 65 °C for 5 min and placing on ice.
Lane 8: 10 μL diluted PCR mixture.
Lane 9: 10 μL diluted PCR mixture.
Lane 10: 10 μL PCR FL-labeled 1.4 kb probe.

Electrophorese for 1–2 h at 50–60 VDC. View the gel under UV light before staining. Stain the gel with ethidium bromide and destain with $H_2O$. View and photograph (see gel photo below).

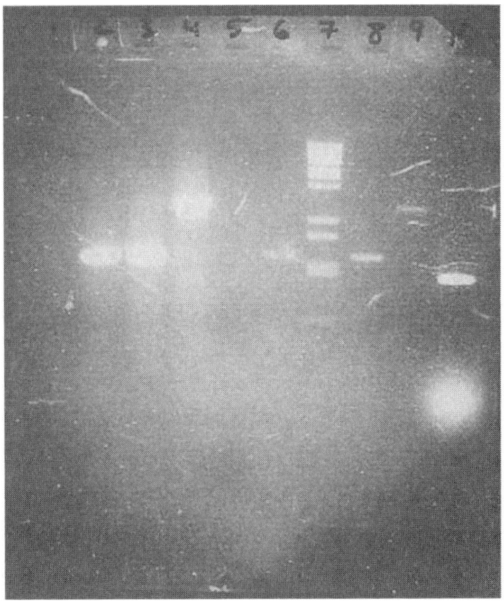

164. What is the large "blob" in the probe lane (lane 10)?

a. dimerized 1.4 kb probe
b. oligomerized 1.4 kb probe
c. polymerized fluorescein
d. fluorescein
e. unincorporated FL-dUTP

165. True or false: After destaining, the gels contain insignificant amounts of ethidium bromide and may be discarded as regular biohazardous trash.

Step 22. To subclone the insert to a plasmid, the following procedure is used: To 50–100 ng of the PCR product from Step 20, above, add $ddH_2O$ and adjust the volume to 4 μL. Add all 4 μL to a tube containing 1 μL (25 ng) of linear pDrive vector and then add to this mixture 5 μL 2X Ligation Master Mix, which contains the ligase buffer and the ligase itself. Incubate the mixture for 0.5–2.0 h at 12 °C. Store at –20 °C.

166. Why is the pDrive plasmid used instead of a λ vector?

a. Plasmids are more stable.
b. Plasmids make the 1.4 kb clone a larger percentage of the total DNA than it would be in a λ vector.
c. Plasmid DNA isolation is easier and more efficient than λ vector DNA.
d. λ is a difficult substrate for sequencing while plasmids are excellent for sequencing.
e. all of the above

167. Why does the 1.4 kb insert, which is really 1.5 kb because the PCR reaction added about 100 bp from the λ phage due to the choice of lambda vector primers ($T_3$ and $T_7$), ligate easily to the pDrive?

a. The single nucleotide "A-overhang" on both ends of the PCR product enables it to readily ligate to the plasmid, which is supplied already cut with a single "T overhang" at each end.
b. The pDrive, which is circular, opens up in the presence of the insert.
c. The pDrive, which is circular, is cut by ligase and then the insert is re-ligated by the ligase.
d. The moderate temperature of the reaction opens the circular form of the pDrive, allowing the insert to chemically bond in the presence of the ligase.
e. The complementarity of the G's at the cut end of the pDrive and C's on the ends of the insert makes for an easy annealing followed by ligation.

Step 23. To 50 μL of competent *E. coli* cells add 2 μL of the pDrive/Rvt ligated DNA. Mix gently. Leave the tube on ice for 5 min and then immediately heat shock by placing the sample in a 42 ºC waterbath for exactly 30 sec. Remove the tube from the waterbath and immediately.place on ice. Add 250 μL of SOC medium (20 g bacto tryptone, 5 g yeast extract, 0.5 g NaCl, 10 mL 250 mM KCl, and $ddH_2O$ to 1 L); autoclave, cool, and add 10 mL sterile $MgCl_2$ and 20 mL 1 M glucose medium to the tube. Prepare an LBA (Luria Broth with ampicillin) plate with X-gal. Spread the cells onto the LBA plate with a sterile glass or plastic cell spreader, allow the plate to sit at room temperature for 10–15 min, invert the plate, and allow it to incubate overnight at 37 ºC. Remove the plate from the 37 ºC incubator and cool to 4 ºC. Pick a white colony that is isolated from other colonies and place in a 10 mL conical tube containing 10 mL of the LBA medium. Incubate overnight at 37 ºC.

168. True or false: Heat-shocking the *E. coli* cells is necessary because *E. coli* normally does not take up circular plasmids or linear DNA.

Step 24. Centrifuge the plasmid cells from Step 23, above. Pour off the supernatant. Resuspend the cells in Buffer P1, which is RNase (100 μg/mL in 50 mM Tris-HCl and 10 mM EDTA, pH 8.0). Add to the cell mixture 250 μL Buffer P2, which is 200 mM NaOH and 1% SDS. Finally, add to the cell mixture 350 μL Buffer N3, which contains guanidine HCl. Pellet the cell debris by centrifuging at 16,000 × g for 10 min. Transfer the supernatant to a QIA prep column. Centrifuge for 10,000 × g for 1 min and discard the eluant. Add 750 μL Buffer PE and centrifuge at 10,000 × g for 1 min. Discard the column eluant. Re-centrifuge at 10,000 × g for 1 min to get rid of the last of the ethanol from the Buffer PE and transfer the column to a new collection tube. Add 50 μL of buffer EB (10 mM Tris, pH 8.5). Centrifuge at 10,000 × g for 1 min and recover the plasmid DNA. The above plasmid isolation method is referred to as an "Alkaline Lysis" protocol.

Step 25. Pour a 1% agarose gel. To 10 μL of the plasmid DNA from Step 24, above, add 190 μL $ddH_2O$ (1 : 20 dilution). Mix 5 μL of the diluted plasmid DNA with 3 μL of $ddH_2O$ and 2 μL of 5X "Blue Juice." Also, mix 2 μL of the undiluted plasmid DNA with 6 μL of $ddH_2O$ and 2 μL of 5X "Blue Juice." Load the gel in the following order:

Lane 3: 10 μL of λBstEII DNA (25 ng/μL = 250 ng) that has been mixed with "Blue Juice" and heated for 3–5 min at 65 ºC and placed on ice.
Lane 4: 10 μL undiluted plasmid.
Lane 5: 6 μL λBstEII DNA (25 ng/μL = 150 ng) that has been mixed with "Blue Juice" and heated for 3–5 min at 65 ºC and placed on ice.
Lane 6: 10 μL of 1:20 diluted plasmid.
Lane 7: 2 μL λBstEII DNA (25 ng/μL = 50 ng) that has been mixed with "Blue Juice" and heated for 3–5 min at 65 ºC and placed on ice.

Electrophorese for 1.5 h at 60 VDC. Stain the gel with ethidium bromide

169. True or false: By comparing the intensity of either the undiluted or the diluted plasmid band with the λBstEII ladders, a semi-quantitative estimate of the amount of DNA in the diluted or undiluted sample can be obtained.

170. What is the most intense band (the one farthest down) in the photo?

a. *E. coli* cell wall protein
b. the supercoiled form of the plasmid
c. normal *E. coli* DNA
d. normal *E. coli* RNA fragments
e. normal *E. coli* mRNA

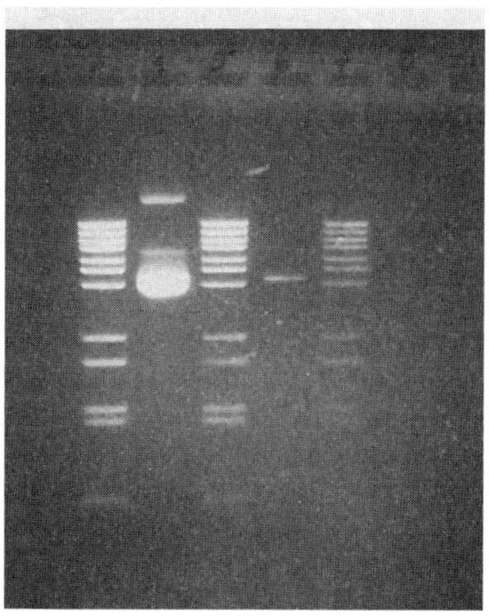

Step 26. To 25–40 ng of the PCR product from Step 21, above, add 4 μL (3.2 pmoles = 20 ng) of T3 sequencing primer and enough $ddH_2O$ to raise the volume to 12 μL. Next, add 8 μL of sequencing reaction pre-mix (buffer, dNTP mix, the four labeled dideoxynucleotides, and *Taq* FS DNA polymerase). The following parameters should be cycled 25 times:

96 °C for 30 sec
50 °C for 15 sec
60 °C for 4 min
and hold at 4 °C after the last cycle

Step 27. Remove the interstitial fluid from a Performa Gel Filtration Cartridge (Edge Bioscience) by centrifuging at 720 × g for exactly 2 min. Discard the eluant and place in a clean collection container on the column. Add the sequencing reaction from Step 26, above, to the top of the spin column. Centrifuge the column at 720 × g for 2 min. Discard the column. Transfer 20 μL of the eluant to a new tube and bring the total volume to 20 μL by adding 10 μL of $ddH_2O$. Sequence on an ABI 3100 (or other) automated DNA sequencer. Compare the sequence obtained to data found at to the National Center for Biotechnology Information (NCBI) by BLAST analysis.

171. Referring to the sequence profile on the inside back cover of the book, which of the following is *true*?

a. All (1–616) of the nucleotides in the sequence are valid and usable.
b. Only the nucleotides below approximately 550 are usable.
c. Only the nucleotides above approximately 50 are usable.
d. Only the nucleotides between approximately 50 and 550 are usable.
e. Only the nucleotides between approximately 150 and 242 are usable.

172. Which of the following nucleotide excision repair (NER) defects would be expected to produce a greatly increased frequency of sunlight-induced skin cancers?

a. Cockayne syndrome (CS)
b. xeroderma pigmentosum (XP)
c. trichothiodystrophy (TTD)
d. all of the above
e. only *a* and *c*

173. Which of the following is a highly conserved DNA sequence in telomeres of such diverse species as birds, reptiles, and humans?

a. TATATA
b. TTAGGG
c. AAAAAA
d. GCGCGC
e. GTTGTT

174. True or false: Angelman syndrome (AS) may result from trisomy 15 (two paternal and one maternal chromosome 15) followed by trisomy rescue in which the maternal chromosome is lost (about 3% of AS cases).

**175–178. Li-Fraumeni Syndrome**

175. Alteration of which of the following genes and its encoded protein are the most frequently encountered genetic events in sporadically occurring human malignancy?

a. *p16*
b. *BRCA1*
c. *BRCA2*
d. HER2/*neu*
e. *p53*

176. True or false: The vast majority of *p53* mutations found in the germline are missense in nature.

177. Which of the following DNA tumor virus genes encode proteins that bind to, and thus inactivate, *p53*?

a. SV40T antigen
b. *E1B* gene of adenovirus
c. the E6 oncoprotein of human papillomavirus (HPV)
d. all of the above
e. only *a* and *b*

178. Which of the following is *true* about the human gene for *p53*?

a. It is located on the short arm of human chromosome 17 band 13.
b. It is approximately 20 kb in length.
c. It has 11 exons, only one of which is non-coding.
d. all of the above
e. only *a* and *b*

179. Which of the following is a tenet of the Lyon (single X) hypothesis (1961)?

a. One of the two X chromosomes in each cell of a female is transcriptionally inactive.
b. Inactivation occurs early in embryogenesis, at which time random inactivation of either a maternally and/or a paternally derived X chromosome occurs in each cell.
c. The random inactivation of the lineage of an X is stable and transmitted to all descendant cells.
d. all of the above
e. only *a* and *b*

**180–182. HNPCC**

180. Which of the following genes is associated with hereditary non-polyposis colon cancer?

a. *hMSH2*
b. *hMLH1*
c. *hPMS1*
d. all of the above
e. only *b* and *c*

181. Which of the following proteins that can mis-match repair heteroduplexes has a particular ability to recognize and bind to "loop outs"?

a. hMUTSα
b. hMUTSβ
c. GT binding protein
d. hMUTSγ
e. hMSH2

182. Of the following consensus markers, which is suitable to determine the presence of microsatellite instability (MSI), is a polyadenine sequence?

a. BAT 25
b. D5S346
c. D2S123
d. D17S250
e. all of the above

183. What is *true* about Q-banding in human chromosomes?

a. It was produced originally with a quinacrine mustard, which is fluorescent.
b. It is primarily adenine-thymidine (A-T) specific.
c. It also can be produced with other dyes such as Hoechst 33258 and DAPI (4,6-diamino-2-phenylindole).
d. It shows brilliant fluorescence in the distal half of Yq.
e. all of the above

184. Which of the following is *true* about the Wobble position in tRNA-mRNA pairing?

a. It probably will be at the 5′ position of the anti-codon.
b. It probably will be at the 5′ position of the codon.
c. It can be at any of the three anti-codon-codon nucleotide pairing positions in bacteria or eucaryotes.
d. It can accept C pairing with U in a eucaryote.
e. It can accept C pairing with U in bacteria.

**185–191. Retinoblastoma**

185. Approximately what percentage of cases of retinoblastoma are unilateral and non-hereditary?

a. 0.1%
b. 1%
c. 10%
d. 60%
e. 99%

186. True or false: The defect in retinoblastoma appears to be the *RB1* locus on chromosome 13.

187. True or false: Since the *Rb* gene has a small CpG island that encompasses the promoter region, aberrant methylation can act as the functional equivalent of a loss-of-function mutation for *Rb*.

188. What is *true* about the protein encoded by the *Rb1* gene?

a. It has a leucine zipper as indicated by the presence of periodic leucine residues every seventh position in an alpha helix.
b. It has a proline-rich region such as observed in the nuclear oncogenes *c-myc* and *c-myb*.
c. It has a full-length sequence of approximately 110 kDa.
d. It has 10 dispersed potential sites for glycosylation.
e. all of the above

189. In the cell cycle

a. the ratio of $p110^{RB}$ to $pp112\text{-}114^{RB}$ is highest in the G1 phase.
b. the production of $pp112\text{-}114^{RB}$ corresponds to the incorporation of thymidine (radioactive) into cellular DNA.
c. the production of $pp112\text{-}114^{RB}$ from $p110^{RB}$ is due to phosphorylation of $p110^{RB}$.
d. all of the above
e. only *a* and *c*

190. pRB can be inactivated by what?

a. mutation
b. deletions
c. methylation
d. viral sequestration, e.g., SV40 and HPV-16
e. any of the above

191. True or false: The p16/Rb pathway is inactivated in both small cell lung carcinoma (SCLC) and non-small cell lung carcinoma (NSCLC) at a very high frequency. However, the targets of the inactivating mutational events are distinct depending on the subtype of the lung tumor analyzed.

**192–196. Neurofibromatosis**

192. Why do two major alternative forms of the NF2 protein exist?

a. because of protein produced from exons 1–15 and exon 17 (isoform 1) and protein produced by the presence of alternatively spliced exon 16 that alters the C-terminus of the protein, replacing 16 amino acids with 11 novel residues in isoform 2
b. because of protein produced from exons 1–17 and a truncated protein produced from only exons 1–15 (isoform 2)
c. because of post-translational modification of the isoform 1 (NF2) to isoform 2 (NF1)
d. because of post-translational modification of the isoform 1 (NF2) to isoform 2 (NF3)
e. because of post-translational modification of the isoform 1 (NF2) to isoform 2 (moesin)

193. Of the following ERM (ezrin-radixin-moesin) proteins, which one is involved in hereditary elliptocytosis?

a. ezrin
b. radixin
c. protein 4.1
d. NF2
e. moesin

194. What is Von Rechlinghausen neurofibromatosis?

a. NF2
b. central fibromatosis
c. NF1
d. bilateral vestibular neurofibromatosis
e. Riccardi-Crowe syndrome

195. In individuals who have reached adulthood and have received a careful examination, including a slit-lamp examination, by an experienced physician, what is the penetrance of neurofibromatosis 1?

a. essentially 100%
b. approximately 50%
c. approximately 10%
d. approximately 1%
e. only about 0.1%.

196. True or false: The *NF1* and *NF2* genes are both on chromosome 17q approximately 0.4cM apart.

197. When treating a renal cell carcinoma cell line in which the *VHL* (von Hippel-Lindau) gene is silent, which of the following agents will cause the *VHL* gene to be re-expressed?

a. bromodeoxyuridine (BrdU)
b. 5-fluorouracil
c. 5-aza-2′-deoxycytidine
d. phytohemagglutinin
e. ubiquinone

198. Which of the following inherited diseases results from mitochondrial inheritance?

a. cystic fibrosis
b. Leber's hereditary optic neuropathy
c. Fragile X syndrome
d. Huntington disease
e. Charcot-Marie-Tooth disease, Type 1A

199. Which of the following HFE (OMIM 602421; Locus ID 1080) genotypes might be highly likely to produce iron overload?

a. wt/C282Y
b. wt/H63D
c. C282Y/C282Y
d. H63D/H63D
e. C282Y/H63D

200. Through methylation and subsequent deamination, cytosine might be converted into which of the following nucleotides?

a. thymine
b. guanine
c. dihydrouridine
d. adenosine
e. pseudouridine

**201–211. Match the following.**

201. ______ 50%
202. ______ 24%
203. ______ 0.1%
204. ______ $2 \times 10^{-9}$ m
205. ______ $3.4 \times 10^{-9}$ m
206. ______ $7.1 \times 10^{-12}$ g
207. ______ UAA
208. ______ E
209. ______ UGG
210. ______ N
211. ______ 22

a. the weight of DNA in a single human cell
b. the amount of the human genome in repeat sequences
c. the amount of the human genome in introns
d. the number of chromosomes excluding X and Y in a haploid cell
e. the width of a DNA double-helix molecule
f. the difference in DNA from one person to another
g. methionine
h. stop codon
i. tryptophan
j. glutamic acid
k. asparagine
l. phenylalanine
m. the distance between base pairs in a single strand of DNA
n. the mass of a single nucleotide
o. the number of nucleotides available to form DNA
p. the distance required for the double helix to rotate 360° every 10 bp

212. Which of the following is/are *true* about a two-step, nested PCR?

a. After the first round of amplification, the initial reaction is stopped and an aliquot of the amplicon is transferred to a second tube. New enzyme and a second set of primers that encompass only a short stretch of the initial amplicon are then added to this tube.
b. A greater expense than straight PCR or one-step nested PCR is incurred due to the use of more enzyme.
c. Sensitivity is increased in many cases.
d. Specificity usually is increased.
e. All of the above are true.

**Questions 213–229 relate to the following scenario:**

Hepatitis C virus (HCV) is the major cause of post-transfusion non-A, non-B hepatitis. HCV is an RNA virus; that is, its genome is RNA rather than DNA. There are five closely related HCV genotypes, with their relative abundance the subject of intensive epidemiologic study. However, genomic sequences located in the 5′-UTR are remarkably well conserved and provide an excellent site for oligonucleotide probes and primers. Detection of HCV infection is accomplished by an RT-PCR assay that consists of the following steps: (1) specimen collection; (2) RNA extraction; (3) reverse transcription; (4) PCR; (5) gel electrophoresis.

213. RNA can be isolated from either serum or plasma (non-heparin anticoagulant) for use in this assay. If the assay cannot be performed immediately, how should the specimen be treated?

a. Store whole blood at room temperature.
b. Store separated plasma/serum at room temperature.
c. Store whole blood at 37 °C.
d. Store separated plasma/serum at –20 °C.

214. A common agent in the isolation of RNA is guanidine thiocyanate (GTC); what is the purpose of GTC in the extraction of RNA?

a. GTC is a chaotropic agent that disrupts tissues and proteins such as RNases.
b. GTC precipitates RNA, which can then be collected by centrifugation.
c. GTC acts with DEPC to degrade contaminating DNA.
d. GTC promotes the partitioning of RNA into the organic phase during PCI extraction.

215. Solutions used in RNA work are typically treated with DEPC. Why?

a. DEPC promotes the precipitation and thus the purification of RNA.
b. DEPC causes single-stranded RNA to fold into double-stranded structures that are more easily isolated.
c. DEPC treatment of solutions destroys contaminating RNases, thus increasing the quantity and quality of the isolated RNA.
d. DEPC degrades contaminating proteins during the initial tissue extraction.

216. Precipitation of isolated RNA is readily accomplished by the addition of which of the following?

a. iso-propanol or ethanol
b. vanadium sulfate
c. chloroform
d. guanidine thiocyanate

217. What is cDNA?

a. double-stranded DNA in which the two strands are oriented antiparallel to each other
b. double-stranded DNA in which the two strands are oriented parallel to each other
c. single-stranded DNA complementary to a single strand of RNA
d. single-stranded RNA complementary to a single strand of DNA

218. Synthesis of cDNA is typically accomplished via the use of which of the following enzymes?

a. *Taq* DNA polymerase
b. Klenow fragment of DNA polymerase
c. DNA ligase
d. avian myoblastosis virus reverse transcriptase

219. Following is the sequence of the oligonucleotide used in cDNA synthesis in a particular experiment: 5′-CCC AAC ACT CGG CTA G-3′. Indicate the mRNA sequence to which this oligonucleotide will anneal in this assay.

a. CCC AAC ACT CGG CTA G
b. GAT CGG CTC ACA ACC C
c. CTA GCC GAG TGT TGG G
d. CUA GCC GAG UGU UGG G

220. Synthesis of cDNA requires all but which of the following?

a. primer
b. NTPs
c. dNTPs
d. $MgCl_2$

221. Each of the following types of primers can be used in cDNA synthesis. Which one is the *least* specific for HCV?

a. oligo-dT
b. random octamers
c. CCC AAC ACT CGG CTA G
d. none of the above

222. Why is RNasin, a recombinant RNase inhibitor, typically added to cDNA synthesis reactions?

a. to degrade contaminating RNA
b. to degrade template RNA following cDNA synthesis
c. to inhibit the action of contaminating RNases
d. to promote the formation of RNA:DNA hybrids that are much more stable to digestion

223. You are setting up a PCR reaction. Which reaction component do you typically add *last*?

a. dNTPs
b. *Taq* polymerase
c. 10X buffer
d. upstream primer

224. Your protocol calls for the addition of downstream primer to a concentration of 1.0 μM. Your primer stock is at a concentration of 10 pmol/μL. If your total reaction volume is 100 μL, how much primer do you need to add?

a. 1.00 μL
b. 5.00 μL
c. 10.00 μL
d. insufficient information is given to determine

225. You are evaluating a PCR method and are trying various concentrations of $Mg^{2+}$ in order to determine the optimum sensitivity of the reaction. You want to set up six separate reactions, where the only parameter that varies is the $MgCl_2$ concentration. Which of the following reaction components could you include in a "master mix" during setup?

a. $MgCl_2$
b. *Taq* polymerase
c. both of the above
d. none of the above

226. Which of the following reagents might you want to include in the PCR reaction in order to reduce reaction-to-reaction carryover?

a. *Taq* polymerase
b. RNasin
c. $MgCl_2$
d. dUTP

227. Following PCR, the assay protocol calls for you to characterize the reaction by gel electrophoresis. If the expected band is 278 bp, which type of gel would you likely want to run to yield optimum results?

a. 0.8% agarose
b. 2.5% agarose
c. 8% polyacrylamide with 8M urea
d. 15% polyacrylamide

228. Following electrophoresis, your patient lane shows a band at 278 bp. Interpret these results.

a. The patient is negative for hepatitis C.
b. The patient is positive for hepatitis C.
c. The results are equivocal. The patient needs to be retested in two weeks.
d. unable to interpret given this limited information

229. As with any diagnostic assay, proper controls are necessary to ensure the functioning of *all* steps of the RT-PCR procedure. Which of the following would be the *best* positive control for the hepatitis C detection assay?

a. RNA from a stock strain of HCV
b. cDNA from a stock strain of HCV
c. Serum from an individual serologically positive for HCV
d. Plasma from an individual uninfected with HCV
e. Serum from a sample known to contain HCV

**230–235. Match the following.**

230. Transcription _____
231. Ethidium bromide _____
232. Intron _____
233. Hybridization _____
234. MOPS buffer, pH 7.0 _____
235. B-form _____

a. fluorescent detection of DNA.
b. synthesis of RNA using a DNA template
c. most common form of DNA in solution
d. untranslated DNA sequence within a gene
e. annealing of probe to template
f. RNA electrophoresis

**236–245. Match the following enzymes to their description or function.**

236. Klenow fragment _____
237. Sequenase _____
238. SP6 RNA polymerase _____
239. Micrococcal nuclease _____
240. MMLV reverse transcriptase ____
241. DNase I _____
242. T4 DNA ligase _____
243. Uracil DNA glycosylase _____
244. Taq polymerase _____
245. Alkaline phosphatase _____

a. used in the PCR method
b. used to prevent contaminant carryover in PCR
c. catalyzes the formation of phosphodiester bonds between nucleic acids
d. reads an RNA strand and synthesizes a complementary DNA strand
e. cleaves phosphodiester bonds in DNA yielding a 5′ phosphate group
f. cleaves phosphodiester bonds in DNA yielding a 3′ phosphate group
g. genetically engineered version of T7 DNA polymerase
h. catalyzes the removal of a 5′ phosphate group
i. used in random primer labeling of DNA
j. used to produce RNA from DNA

246. The nucleic acid based sequence amplification (NASBA) technique utilizes which of the following enzyme *activities*?

1. T4 polynucleotide kinase
2. RNase H
3. DNA ligase
4. reverse transcriptase

a. 1, 2, and 3 are correct.
b. 1 and 3 only are correct.
c. 2 and 4 only are correct.
d. 4 only is correct.

247. What would be the effect if you forgot to add dNTPs to your ligase chain reaction?

1. There would be no amplification.
2. Linear amplification only would occur.
3. Exponential amplification only would occur.
4. There would be no effect; LCR does not need dNTPs.

a. 1, 2, and 3 are correct.
b. 1 and 3 only are correct.
c. 2 and 4 only are correct.
d. 4 only is correct.

248. Target amplification reactions include which of the following?

1. hybrid capture amplification
2. strand displacement amplification
3. branched DNA amplification
4. RT-PCR

a. 1, 2, and 3 are correct.
b. 1 and 3 only are correct.
c. 2 and 4 only are correct.
d. 4 only is correct.

249. Which of the following enzymes/enzyme combinations are used to prepare strand-specific probes?

1. Klenow fragment of DNA polymerase DNase I
2. T3 RNA polymerase
3. DNA polymerase I and DNase I
4. SP6 RNA polymerase

a. 1, 2, and 3 are correct.
b. 1 and 3 only are correct.
c. 2 and 4 only are correct.
d. 4 only is correct.

250. Nick translation labeling can be accomplished with which labeled nucleotide?

1. $\alpha$-[$^{32}$P]-UTP
2. $\gamma$-[$^{32}$P]-dATP
3. digoxigenin-11-UTP
4. biotin-16-dUTP

a. 1, 2, and 3 are correct.
b. 1 and 3 only are correct.
c. 2 and 4 only are correct.
d. 4 only is correct.

251. T4 polynucleotide kinase is used in which reaction?

a. 5′ end labeling
b. allele-specific PCR
c. multiplex PCR
d. branched DNA amplification

**252–261. Match the gene/translocation with the disease/protein type.**

252. *c-MYC* with immunoglobulin loci/ t(8;14) (q24;q32) or t(2;8) (p12;q24) or t(8;22) (q24;q11) _____
253. *c-ABL* (9q34) with *BCR* (22q11)/t(9;22) (q34;q11) _____
254. *BCL-2* with immunoglobulin loci/t(14;18) (q32;21) _____
255. *BCL-3* with immunoglobulin loci/t(14;19) (q32;q13.1) _____
256. *PML* (15q21) with *RAR*(17q21)/t(15;17) (q21;q11-22) _____
257. *REL* with *NRG*/ins(2;12) (p13;p11.2-14) _____
258. *PBX-1* (1q23) with *E2A* (19p13.3)/t(1;19) (q23;p13.3) _____
259. *HOX-11* with T-cell receptor loci/t(10;14) (q24;q11) t(7;10) (q35;q24) _____
260. *SSX1*, *SSX2*, *SYT* t(X;18) (p11.2;q11.2) _____
261. *ERG* with *EWS*/t(21;22) (q22;q12) _____

a. Ewing sarcoma/Ets transcription factor family
b. acute pre-B cell leukemia/homeodoman
c. chronic B-cell leukemia/CDC10 motif
d. chronic mylogenous leukemia and acute leukemia/tyrosine kinase activated by B-cell receptor
e. synovial sarcoma/helix-loop-helix domain
f. B-cell chronic lymphocyte leukemia/PRADI-G1 cyclin
g. acute pre-B-cell leukemia/homeodomain helix-loop-helix
h. lipomas/HMG DNA-binding protein
i. non-Hodgkin lymphoma/NF-κB family
j. Burkitt lymphoma, BL-ALL/helix-loop-helix domain
k. follicular lymphoma/inner mitochondrial membrane
l. acute myeloid leukemia/zinc finger
m. acute T-cell leukemia/homeodomain

**262–280.** An experiment to construct one mouse liver cDNA library starts with poly (A)+ mRNA (5′-NNNNNNNNNAAAAAA-3′).

Step 1. Poly (A)+ mRNA that has been precipitated with ethanol is centrifuged at high speed for 60 min. The ethanol supernatant is removed with a micro pipette without disturbing the pellet. The pellet is washed with ice-cold 70% ethanol; the poly (A)+ mRNA is re-pelleted with centrifugation. The excess ethanol is removed, and the tube is allowed to air-dry for 5 min.

262. True or false: The mRNA at this point should have the intron sequences spliced out.

Step 2. To the mRNA is added 37.5 μL of DEPC (diethylpyrocarbonate)-treated water that has been autoclaved and 1.0 μL of RNase inhibitor (40 U/μL). The pellet is softened by allowing it to stand for 5 min on ice. To this tube is now added the following mixture:

5.0 μL 10X first-strand buffer
3.0 μL methylated dNTP mix
2.0 μL linker-primer, 1.4 μg/μL

The linker-primer is 5′-$(GA)_{10}$ACTAGT**CTCGAG**$(T)_{13}$-3′

263. What is the purpose of the DEPC-treated water?

a. to provide water that is free from RNase
b. to provide an inhibitor to reverse transcriptase until the actual reverse transcriptase is added
c. to provide the proper environment for the annealing process
d. to keep the methylated dNTPs from demethylating.
e. to demethylate the dNTPs

264. Where does the linker-primer anneal to the mRNA?

a. at a series of AGs on the mRNA
b. at the UGAUCAGCUC motif in the mRNA
c. at the poly-A tail in the mRNA
d. at the ACUAGUCUCGAG motif in the mRNA
e. at the GAGCUCUGAUCA motif in the mRNA.

Step 3. The tube containing the reaction mixture is allowed to stand at room temperature for 10 min. After the 10-min period, 1.5 μL of reverse transcriptase (50 U/μL) is added and the contents of the tube are mixed and incubated at 37 °C for 1 h.

265. True or false: The first-strand cDNA that is synthesized is non-methylated as the reverse transcriptase accomplishes both synthesis of cDNA and demethylation contemporaneously.

Step 4. After completion of first-strand synthesis, the tube is centrifuged briefly and placed on ice. Next the following components are added in the order stated:

20.0 μL 10X second-strand buffer
6.0 μL dNTPs
116.0 μL of DEPC-treated $H_2O$

Mix, centrifuge, and then add

2.0 μL RNase (1.5 U/μL)
11.0 μL DNA Polymerase I (9 U/μL)

Incubate at 16.0 °C for 2.5 h.

266. What occurs during the above series of reactions?

a. the original RNA template is removed and a second strand of DNA complementary to the originally synthesized cDNA [except for the addition of 5′-$(GA)_{10}$ACTAGTCTCGAG$(T)_{13}$-3′ to the poly-A tail] is produced with the loss of about 5–15 nucleotides in the 5′-UTR.
b. the original RNA template is removed and a second strand of DNA complementary to the originally synthesized cDNA is produced.
c. the original RNA template is removed and a second strand of DNA complementary to the originally synthesized cDNA is produced with the loss of about 5–15 nucleotides in the 5′-UTR and the addition of 5′-$(GA)_{10}$ACTAGTCTCGAG$(T)_{13}$-3′ to the poly-A tail.
d. all of the above occur, giving rise to a mixture.
e. the original RNA template is removed and a second strand of DNA complementary to the originally synthesized cDNA is produced with methylated nucleotides and with the loss of about 5–15 nucleotides in the 5′-UTR and the addition of 5′-$(GA)_{10}$ACTAGTCTCGAG$(T)_{13}$-3′ to the poly-A tail.

Step 5. To the second-strand mixture add the following ice-cold components while the mixture is maintained on ice:

23.0 μL Blunting dNTP mix
2.0 μL *Pfu* polymerase (2 U/μL)

Mix, centrifuge, and then place in a 72 °C heated block for 30 min. Place on ice immediately thereafter.

267. True or false: At this point, the two complementary strands of DNA are of equal length.

Step 6. Transfer the reaction mixture from step 5, above, to a pre-centrifuged phase lock tube (gel barrier tube). To the phase lock gel tube add 200 μL of buffered phenol and vortex. Next add 200 μL of chisam, which is 24:1 chloroform/iso-amyl alcohol and vortex again.

Step 7. Remove the upper aqueous layer from the tube in step 6, above, and transfer to a new tube. To the new tube add 20 μL of 3M sodium acetate (pH 5.2), close the tube, mix briefly, and then add 400 μL of ice-cold 100% ethanol. Invert several times and precipitate the cDNA at –20 °C.

Step 8. Centrifuge the precipitated cDNA, remove the supernatant ethanol with a micropipette, and wash the pellet with 500 μL of 70% ethanol. Remove the ethanol and allow the pellet to air dry.

Step 9. Pipette 9.0 μL of EcoRI adaptors (5′-HO-AATTCGGCACGAG-3′/3′-GCCGTGCTC-P-5′) into the tube containing the dry pellet. Resuspend the pellet. Next add the following components:

1.0 μL 10X ligase buffer
1.0 μL rATP (regular ATP as opposed to dATP for DNA synthesis)
1.0 μL T4 DNA ligase (4 U/μL)

Mix and incubate at 8 °C overnight.

268. Which of the following is *true* about this point in the reaction sequence?

a. The second strand of cDNA has a dephospho-AATTCGGCACGAG on the 5′ end of the original (methylated) cDNA strand and a GCCGTGCTC-dephospho-3′ on the other end.
b. Both the second strand of cDNA (5′ end) and the original, methylated strand of cDNA (3′ end) have an AATT overhang.
c. The second strand of cDNA has a 5′-$PO_4$-GCCGTGCTC and a AATTCG-GCACGAG-$PO_4$ on the 3′ while the original (methylated) cDNA strand has the opposite sequence.
d. The second strand of cDNA has a $PO_4$-AATTCGGCACGAG on the 5′ end and a GCCGTGCTC-$PO_4$ on the other end while the original (methylated) cDNA strand has the opposite additions.
e. Only the second strand of cDNA has a $PO_4$-AATTCGGCACGAG on the 5′ end and a GCCGTGCTC-$PO_4$ on the other end.

Step 10. Stop the reaction by heating at 70 °C for 30 min and briefly centrifuge.

Step 11. To the reaction mixture from Step 10, above, add the following components:

1.0 μL 10X LIGASE buffer
2.0 μL rATP (10 mM)
5.0 μL $H_2O$
2.0 μL T4 polynucleotide kinase (10 U/μL)

Mix the reaction components and incubate at 37 °C for 30 min.

269. True or false: At this point in the reaction sequence all four ends of the dsDNA are phosphorylated.

Step 12. Stop the reaction by heating at 70 °C for 30 min.

270. In steps 10 and 12, above, the reactions were stopped using what?

a. hyperphosphorylation of the enzyme
b. melting of the dsDNA
c. removal of the end nucleotides with heat
d. heat inactivation of the enzyme involved
e. reversible heat-induced demethylation of the original strand of DNA.

Step 13. To the mixture from Step 12, above,

- add 28 μL of XhoI buffer supplement and mix;
- add 3.0 μL XhoI, mix, and incubate for 1.5 h at 37 °C;
- centrifuge and cool for 5 min.

271. What is the purpose of adding the XhoI?

a. to remove all of the terminal phosphates
b. to remove only the terminal phosphates on the second strand 5′ end and the original strand 3′ end
c. to cut the 5′-CTCGAG-3′/3′-GAGCTC-5′ ds cDNA sequences at the CT pairs on the poly A tail end of the cDNA right arm creating a unique "Xho end"
d. to remove the methyl groups on the first strand
e. to cut the 5′-CTCGAG-3′/3′-GAGCTC-5′ ds cDNA sequences at the CG pairs creating a unique "Xho end"

272. True or false: The first strand and second strand of cDNA are not cut by XhoI (except at the position of the original linker primer) since the first strand of cDNA is methylated.

Step 14. Prepare a Chromaspin-400 column. To the ds cDNA add 0.1% bromophenol blue (not "Blue Juice") and mix. Add the double-stranded cDNA with bromophenol blue onto the middle of the Chromaspin column. Centrifuge for 5 min at 700 × g. Save the eluate and bring it to a total volume of about 100 μL.

273. What will the eluate from the Chromaspin-400 column contain?

a. ssDNA (the original strand and the second strand as a mixture)
b. small pieces of DNA from the XhoI digestion
c. degraded DNA
d. small cDNAs made from degraded mRNA templates
e. ds cDNA that will be inserted into the lambda vector

Step 15. Pipette all of the eluate into a pre-spun phase lock tube (gel barrier tube). Add 50 μL of buffered phenol to the mixture in the phase lock tube and vortex for 30 sec. Next, add 50 μL chisam (chloroform/*iso*-amyl alcohol; 24:1) and vortex for another 30 sec. Centrifuge for 2 min.

274. The chisam in step 15, above, is used to what?

a. to remove small organic molecules
b. to remove salts
c. to remove short oligos
d. to extract proteins and degraded proteins from the aqueous solution that contains the DNA
e. only *a*, *b*, and *c*

Step 16. Remove the upper layer from the phase lock tube and place on a 45 mm Millipore membrane (VSWP 04500, pore size: 0.025 μm, type VS) that has been floated shiny side up on about 30 mL of sterile $H_2O$. Allow dialysis to proceed for about 5–6 h.

275. What is the upper layer?

a. the aqueous layer
b. $CHCl_3$
c. $CHCl_3$/phenol
d. $CHCl_3$/phenol/*iso*-amyl alcohol
e. the gel layer

276. What is the purpose of the dialysis step?

a. to convert ssDNA into dsDNA by lowering the salt content in the droplet on the membrane
b. to convert dsDNA into ssDNA by pulling water up into the droplet on the membrane
c. to lower the salt content in the droplet when salts diffuse into the water below the membrane
d. to raise the salt content in the droplet on the membrane when water diffuses into the larger volume of water below the membrane
e. to remove oligos, which will diffuse into the water below the membrane

Step 17. Retrieve as much of the dialysate as possible, transfer to a new centrifuge tube, and concentrate to a volume of exactly 5 μL. Remove exactly half (2.5 μL) and transfer to a new tube. To the 2.5 μL of dialysate add the following components:

0.5 μL 10X ligase buffer
0.5 μL rATP (10 mM)
1.0 μL Uni-ZAP XR lambda vector that has been pre-cut with EcoRI and XhoI and dephosphorylated
0.5 μL T4 ligase

Mix the above and incubate at 12 °C overnight.

277. How will the ds cDNA insert into the lambda vector?

a. unidirectionally, because one end of the cDNA has an EcoRI and the other end has an XhoI end
b. in any one of two ways, due to the nature of the ends of the ds cDNA and the cut ends of the vector
c. in any one of four ways, due to the nature of the ends of the ds cDNA and the cut ends of the vector
d. randomly
e. in any one of three ways, due to the nature of the ends of the ds cDNA and the cut ends of the vector

Step 18. To a freshly thawed packaging extract tube that has been kept on dry ice, add all 5.0 μL of the reaction mixture from Step 17, above. Mix and incubate for no longer than 2 h at room temperature. Add 500 μL SM buffer and 20 μL chloroform. Centrifuge to sediment debris. Store at 4 °C.

278. True or false: Although the use of 4 °C is acceptable for preservation of the phage, – 20 °C is preferable.

Step 19. To a microfuge tube containing 200 μL of plating cells (XL-1 Blue MRF), add 1 μL of packaged bacteriophage from step 18, above. Perform the same operation with 1 μL of packaged bacteriophage that has been diluted 1:50 with SM buffer. Incubate both tubes for 15 min at 37 °C.

Step 20. Select two NZY (NaCl; $MgSO_4$; yeast extract; NZ amine, which is casein hydrolysate; and NaOH to pH 7) plates (one undiluted and one diluted).

To a tube of 3 mL of molten (50 °C) top agarose that contains 15 μL of 0.5 M IPTG (isopropyl-β-D-thiogalactopyranoside)

- add 50 μL of 250 mg/mL X-gal (5-bromo-4-chloro-3-indolyl-β-D-galactoside),
- add all 201 μL of the undiluted packaged bacteriophage and cells,
- mix, and
- immediately pour onto a labeled plate.

Repeat the operation in a separate tube with the diluted packaged bacteriophage and cells. Allow the plates to sit at room temperature for 5–10 min to solidify and then incubate at 37 °C inverted overnight.

Step 21. Count the total (blue plus clear) plaques in whichever plate has an optimal number of plaques for counting. Also count the number of clear plaques.

279. True or false: Insert-containing plaques are clear.

280. True or false: Plaques with an intact lacZ gene will be blue.

281. Which of the following is a chromosomal translocation typical of ataxia-telangectasia?

a. 9:22
b. 7:14
c. 8:22
d. 7:11
e. 7:22

**282–289. Match the hemoglobinopathies/thalassemias and their genetic defect.**

282. Hemoglobin lepore _____
283. Hereditary persistence of fetal hemoglobinemia (HPFH) _____
284. $\beta^0$-Thalassemia _____
285. Hemoglobin S _____
286. $\alpha$-Thalassemia-2 trait (rightward) _____
287. Hemoglobin Constant Spring _____
288. Hemoglobin C _____
289. $\alpha$-Thalassemia-2 trait (leftward) _____

a. the result of changing GAG to GTG on codon 6 of the first exon of the $\beta$-chain gene, resulting in the substitution of valine for glutamine
b. deletion of one $\psi$ gene
c. a 3.7-kb deletion from one chromosome, including a 3′-portion of the upstream $\alpha$2-globin gene, a 5′ portion of the downstream $\alpha$1-globin gene, and more than 2000 nucleotides that normally are between the two $\alpha$ genes, with a resulting fusion of the $\alpha$2-gene and part of the $\alpha$1 gene
d. the result of a TAA to CAA mutation that specifies glutamine instead of termination after addition of the $141^{st}$ amino acid on the $\alpha$-chain
e. African type results from (rightward) deletion of the entire $\delta\beta$-gene region
f. $\beta$-chain synthesis is absent and no hemoglobin A is demonstrable
g. a 4.2-kb segment of DNA, including the entire upstream $\alpha$2-globin gene
h. non-homologous meiotic (leftward) crossing-over of the $\delta$- and $\beta$-genes
i. the result of the replacement of glutamine with lysine on the $\beta$-chain

290. A mutation in which of the following genes represents the basic cause of Bloom syndrome?

a. BLM helicase
b. DNA ligase I
c. $O^6$-methylguanine methyltransferase
d. N-methylpurine DNA glycosylase
e. superoxide dimutase

**291–296. *BRCA1/BRCA2***

291. *BRCA1* and *BRCA2* mutations represent approximately what percentage of breast cancers that cluster in families?

a. 95–99%
b. 50–60%
c. 15–20%
d. 5–8%
e. 1%

292. Which of the following is/are a technique used to identify *BRCA1* mutation?

a. protein truncation assay
b. multiplex heteroduplex analysis
c. conformation-sensitive gel electrophoresis
d. direct sequencing
e. any of the above

293. True or false: The two most common mutations in the *BRCA1* gene are 185delAG and 5382insC.

294. True or false: *BRCA1* is a nuclear phosphoprotein that shows greatest expression and phosphorylation in the S and M phases of the cell cycle.

295. Which of the following statements is *true* about *BRCA1* and *BRCA2*?

a. They are both localized on chromosome 17q21-22 and contiguous.
b. They are both localized on chromosome 13q12-13 and contiguous.
c. They are both localized on the short arm of chromosome 17 approximately 0.6 cM apart.
d. They are on separate chromosomes (17q21 for *BRCA1* and 13q12-13 for *BRCA2*) but linked by disease state (i.e., when *BRCA1* is mutated so is *BRCA2*, and vice versa).
e. They are on separate chromosomes (17q21 for *BRCA1* and 13q12-13 for *BRCA2*) but are independent (i.e., when *BRCA1* is mutated, *BRCA2* may not be, and vice versa).

296. True or false: *BRCA2* mutations appear to be spread across the entire gene with no mutation "hot spots."

**297–299. Cervical Cancer**

297. Approximately what percentage of invasive cervical carcinomas has been shown to contain the DNA sequences from particular HPV (Human Papilloma Virus) types?

a. >90%
b. approximately 75%
c. approximately 50%
d. approximately 10%
e. <1%

298. Which of the following HPV genes inactivates human p53?

a. *E1*
b. *E3*
c. *E6*
d. *E7*
e. all of the above

299. Which of the following HPV genes inactivates RB1?

a. *E1*
b. *E2*
c. *E6*
d. *E7*
e. *E16*

**300–302. Werner Syndrome**

300. True or false: In all WRN mutations identified to date and those that result in Werner syndrome, a truncated protein (WRNp) is obtained.

301. True or false: Hutchinson-Gilford syndrome, a progeroid syndrome with childhood onset, has a mutation of the WRNp in which the protein is longer than the wild-type 1432 amino acid WRNp.

302. Although the *in vivo* function of the WRNp is unknown, what is the *in vitro* function of the WRNp?

a. It acts as a cytosine methylase.
b. It acts as a topoisomerase.
c. It acts as a $3' \rightarrow 5'$ DNA helicase.
d. It acts as a striated muscle structural component that interacts synergistically with troponin-T.
e. It acts as a striated muscle structural component that interacts synergistically with troponin-C.

303. Which of the following is the eukaryotic translation start codon?

a. AGU
b. CAU
c. UGA
d. AUG

304. Which statement is *false*?

a. Mitochondria use the same genetic code as bacteria.
b. UAA does not code for an amino acid.
c. Guanine can base pair with both cytosine and uracil in tRNA.
d. The <u>anticodon</u> for Met is CAU.

305. What is a suppressor mutation?

a. a mutation that suppresses expression of a gene
b. a second mutation that suppresses the effects of the first mutation
c. a mutation that causes a tRNA to bind to the P site instead of the A site
d. a mutation that introduces a frameshift into the synthesis of a tRNA

306. In protein maturation, the absence of a signal causes a protein to localize where?

a. the membrane
b. the endoplasmic reticulum
c. It is exported out of the cell.
d. the cytosol

307. Where are protein targeting signals located?

   a. in the N-terminal region
   b. in the C-terminal region
   c. in the internal region
   d. in possibly all of the above regions

308. Which of the following is/are *true* about protein folding?

   a. It may be mediated by chaperone complexes.
   b. It can occur in the absence of outside factors.
   c. Answers *a* and *b* are both correct.
   d. None of the above is correct.

309. Mitochondrial protein transport involves which of the following?

   a. TOM channel
   b. TIM channel
   c. C-terminal trailer sequences
   d. Only *a* and *b* are correct.

310. Nuclear recognition signals have which type of amino acids?

   a. hydrophobic
   b. aromatic
   c. acidic
   d. basic

311. Which protein covalently attaches to proteins and targets them for degradation?

   a. trypsin
   b. ubiquitin
   c. lysozyme
   d. ARC

312. Which of the following transcribes mRNA?

   a. RNA polymerase I
   b. RNA polymerase II
   c. RNA polymerase III
   d. RNA polymerase $\alpha$

313. Which of the following is a consensus–10 sequence?

   a. TATAAT
   b. ATATAT
   c. TGTTTC
   d. AAGAAG

314. What is the number of RNA bases complexed with RNA polymerase II at any given time?

   a. 12
   b. 20
   c. 25
   d. 55

315. What is the rate of transcription of bacterial RNA polymerase at 37 °C?

a. 800 bp/sec
b. 200 bp/sec
c. 100 bp/sec
d. 40 bp/sec

316. What is the sequence of events in recovery from a transcriptional pause?

a. backtracking, cleavage, new catalysis
b. cleavage, backtracking, new catalysis
c. backtracking, new catalysis, cleavage
d. cleavage, new catalysis, backtracking

317. Which of the following is the domain of *E. coli* sigma factor that makes contacts with DNA?

a. N-terminal domain
b. central domain
c. C-terminal domain
d. Both *a* and *c* are correct.

318. Which of the following is *not* a cis-acting regulatory element?

a. N protein
b. –10 sequence
c. $nut_R$ sequence
d. $O_R3$

319. β-galactosidase is coded for by which gene?

a. *lacA*
b. *lacY*
c. *lacZ*
d. *lacI*

320. Which of the following enzymes is used to specifically cut DNA at the site of a cyclobutane pyrimidine dimer?

a. photolyase
b. AP lyase
c. T4 endonuclease V
d. DNA ligase
e. both *a* and *c*

321. In the Form I-Form II assay, why does form II plasmid migrate slower than form I plasmid?

a. Form II plasmid is longer than form I plasmid.
b. Form I plasmid, being tightly supercoiled, has a lower effective area and can thus migrate faster than Form II, which is a nicked open circle.
c. DNA damage causes distortion in the molecule, resulting in faster migration of form I plasmid.
d. Form II plasmid is complexed with protein; the larger complex migrates slower due to increased mass.

322. In eukaryotes, the β-actin gene is "constitutive." What does this tell you about its expression control?

a. Expression of the gene is activated by removal of a repressor.
b. Expression of the gene is repressed by removal of an activator.
c. Gene expression can be induced to high levels.
d. Expression of the gene is not subject to regulation.

323. What is the DNA binding motif common to cI, cro, and lac repressor?

a. leucine zipper motif
b. helix-turn-helix motif
c. zinc finger motif
d. acid blob domain

324. Which of the following is an *antiterminator* that functions during phage lambda infection of *E. coli*?

a. N protein
b. Q protein
c. Cro protein
d. Both *a* & *b* are correct.

**325–329. Wilms Tumor**

325. True or false: Wilms tumor is similar to retinoblastoma in that the hemizygous state is phenotypically silent.

326. Where is the Wilms tumor suppressor gene, *WT1*, located?

a. 11p13
b. 11p15
c. 17q12-21
d. 1p18
e. 17p18

327. Which of the following statements is/are *true* about the protein encoded by the *WT1* gene?

a. It has a C-terminus that contains four cysteine-histidine zinc finger domains.
b. It has an N-terminus that is rich in prolines and glutamines, a feature that is shared by the transactivation domains of some transcription factors.
c. It is 35 kDa.
d. All of the above are true.
e. Only *a* and *b* are true.

328. How is WT1 expressed in humans?

a. at a high in the 20-week kidney and barely detectable in the adult organ
b. in a continuously increasing fashion in the developing kidney with a rapid cessation of expression to barely detectable levels at 35–38 weeks gestation
c. at persistently high levels in adults in specific cells types such as Sertoli cells of the testis, granulosa cells of the ovary, muscle cells of the uterus, stromal cells of the spleen, and mesothelial cells of the pericardial pleural and peritoneal lining
d. in acute leukemia cells
e. *a*, *c*, and *d*

329. Children with specific germ-line mutations that affect the *WT1* DNA-binding domains develop a severe constellation of abnormalities known as what?

a. Beckwith-Wiedemann syndrome
b. Denys-Drash syndrome
c. Duchenne muscular dystrophy
d. Knudson-Strong syndrome
e. Beckwith-Bennington syndrome

**330–334. Mitosis. Match the following.**

| | | |
|---|---|---|
| 330. _____ | G0 | a. DNA duplication |
| 331. _____ | S | b. quiescence |
| 332. _____ | G2 | c. actual cell division |
| 333. _____ | M | d. preparation for mitosis |
| 334. _____ | G1 | e. pachytene |
| | | f. Barr body formation. |
| | | g. preparation for DNA replication |

**335–343. Meiosis. Match the following.**

335. _____ Bivalent
336. _____ Chiasma
337. _____ Leptotene
338. _____ Zygotene
339. _____ Pachytene
340. _____ Diplotene
341. _____ *Spo*11
342. _____ Telophase II
343. _____ Diakinesis

a. where chiasmata are first seen at the end of the prophase. The synaptonemal complex begins to disappear.
b. a crossover between two non-sister chromatids
c. a stage of the prophase where the synaptonemal complex (SC) begins to develop between two sets of sister chromatids
d. the beginning of the prophase where duplicated, paired homologues condense
e. a meiotic endonuclease that induces double-stranded DNA breaks at preferential "hot spots"
f. ends with the formation of nuclear envelopes around the four haploid nuclei produced
g. shortening of chromosomes and the chiasmata appear to move to terminalize
h. when synapsis is complete in the prophase and ends with desynapsis
i. Barr body formation
j. splitting of kinetochores

344. Very early genes in λ virus include all but which of the following?

a. N gene
b. Cro gene
c. cI gene
d. All of these are late genes.

345. Cro protein binds with high affinity to which of the following sites?

a. $O_R3$
b. $O_R1$
c. –35 sequence
d. $O_{lac}$

346. Which of the following histones would be expected to be associated with two superhelical turns?

a. H1, H2, H3, and H4
b. H2A, H2B, H3, and H4
c. H1, $(H2A)_2$, $(H2B)_2$, $(H3)_2$, $(H4)_2$
d. $(H2A)_2$ and $(H2B)_2$
e. Only H1

347. Bloom syndrome exhibits which of the following unique observable laboratory phenomena?

a. an unusually high rate of microscopically observable chromosomal breakage
b. an unusually high rate of sister chromatid exchange (blood lymphocytes in short-term culture)
c. an unusually high rate of UV-induced unscheduled DNA synthesis (UDS) in cultured fibroblasts
d. double minutes
e. both *a* and *b*

348. A full expansion (> 200) of which of the following trinucleotide repeats just 5′ to the FMR gene on the long arm of the X chromosome gives rise to Fragile X syndrome?

a. CAG
b. GTT
c. GAA
d. CTG
e. CGG

349. Which of the following is the Factor V Leiden mutation that leads to genetic susceptibility to thrombosis?

a. the substitution of an adenine for guanosine at nucleotide 1691
b. the replacement of arginine by glutamine at amino acid codon 506
c. $FVR^{506}Q$
d. all of the above
e. only *a* and *c*

**350–363. Translocations. Match the following.**

350. _____ t(8;21)(q22;q22)
351. _____ t(15;17)(q31;q22)
352. _____ t(9;11)(p22;q23)
353. _____ inv(16)
354. _____ t(4;11)(q21;q23)
355. _____ t(1;14)(p34;q11)
356. _____ del 13q14-23.1
357. _____ 11q22.3-q23.1
358. _____ t(8;14)(q24;q32.3)
359. _____ t(11;18)(q21;q21) or t(1;14)(p22;q32)
360. _____ t(14;18)(q24;q32)
361. _____ t(11;14)(q13;q32)
362. _____ t(9;14)(p13;q32)
363. _____ t(2;5)(p23;q35)

a. $CD10^-/CD15^\pm/CD33^\pm/CD65^\pm$ B-cell precursor phenotype of acute lymphoblastic leukemia (ALL); infant and older adult age groups, hyperleukocytosis; dismal outcome
b. affects PML-RARα gene. About 7% of acute myelogenous leukemia (AML)
c. $CD10^-$ T-cell phenotype of ALL; male predominance, hyperleukocytosis; intermediate prognosis
d. a common chromosomal abnormality in chronic lymphocytic leukemia (CLL)
e. anaplastic large cell lymphoma
f. mantle cell lymphoma
g. affects AML1(CBFα)-ETO genes. Present in about 12% of patients with acute mylogenous leukemia (AML); associated with the loss of Y in males or an X in females in over half of cases
h. affects ALL1 (MLL)-AF9 genes. Associated with monocytic leukemia. Rearranges ALL1(MLL) genes
i. the most common chromosomal abnormality in chronic lymphocytic leukemia (CLL)
j. lymphoplasmacytoid
k. MALT lymphoma
l. affects CBFB-MYH11 genes. Present in about 12% of cases of AML; associated with marrow eosinophils
m. follicular lymphoma
n. B-cell phenotype of acute lymphocytic leukemia (ALL); L3 morphology; male predominance; bulky extramedullary disease; favorable prognosis with short-term, intensive chemotherapy including high-dose methotrexate and cyclophosphamide
o. Burkitt lymphoma
p. monsomy 7

**364–366. Colon Cancer**

364. Which of the following is *true* about the wild-type *p53* gene?

a. It is a proto-oncogene.
b. It is an oncogene.
c. It codes for reverse transcriptase.
d. It is a tumor suppressor gene.
e. It codes for a cytosine methylase.

365. The *p53* gene encodes a protein that

a. is a tetramer.
b. has a consensus binding site 5′-PuPuPuC(A/T)(A/T)GPyPyPy-3′.
c. binds to DNA even when only two copies of the protein are present.
d. has arginines at 248 and 273 that are frequent sites for somatic mutation and that bind directly to DNA (T and A, respectively).
e. all of the above

366. A tumor suppressor gene on chromosome 18 might be which of the following?

a. DCC (deleted in colon cancer)
b. DPC4 (deleted in pancreatic cancer; re-designated SMAD4)
c. SMAD2
d. all of the above
e. only *b* and *c*

**367–389.** The purpose of the following experiment is to isolate and purify mouse genomic DNA. The sample will be digested with different restriction endonucleases and the way in which methylation influences digestion will be investigated. DNA samples will be transferred to a nylon membrane. The organization of the mouse 1.4 kb repeat containing the reverse transcriptase gene will be investigated and the *Ttr* (transthyretin) gene will be detected in PCR products.

Step 1. Frozen (liquid nitrogen or $LN_2$) mouse liver pieces are pulverized under $LN_2$ in a mortar and pestle. The pulverized tissue suspended in $LN_2$ is poured into a conical plastic tube and the $LN_2$ is allowed to evaporate. To the dried pellet is added Qiagen Buffer G2, which contains RNase, Triton X-100, and proteinase K.

367. What is the purpose of the proteinase K in step 1, above?

a. to digest proteins that are present in the mouse liver sample
b. to add a potassium ion to each from carboxylic acid group (including the carboxy terminus) of all proteins present
c. to protect any histones that might be associated with the genomic DNA
d. to disassociate any histones from the genomic DNA for reassembly in step 4
e. only to disrupt the quarternary structure of the histones associated with the DNA

368. What is the purpose of the RNase in step 1?

a. to protect RNA that might be associated with the genomic DNA
b. to disassociate RNA from genomic DNA for later re-association in step 4
c. to digest RNA that is present in the mouse liver sample
d. to remove any radionucleotides that might be present in the genomic DNA
e. to disassociate any histones from RNA

369. What is the purpose of the Triton X-100?

a. to hydrate the mixture
b. to dehydrate the mixture
c. to dehydrate the pellet completely
d. to coat the walls of the tube in which the reaction is taking place
e. to provide detergent for solubilization of pellet components (particularly membranes)

Step 2. The mixture from step 1, above, is allowed to incubate overnight at 50 °C with gentle rocking. The digest mixture in the conical tube is then centrifuged at 5,000 × g for 10 min. The supernatant is transferred by careful pipetting into a new conical tube without disturbing the sediment in the bottom, vortexed at maximum speed for exactly 10 sec, and poured into a Qiagen Genomic-tip column that has been equilibrated with Buffer QBT (750 mM NaCl; 50 mM MOPS buffer, pH 7.0; 15% isopropanol; and 0.15% Triton X-100). The supernatant is allowed to pass through the column using gravity flow, and the eluate is discarded. After the supernatant has passed through the column completely, the column is washed with Buffer QC (1 M NaCl; 50 mM MOPS, pH 7.0; and 15% isopropanol). The column eluate is again discarded. Using a glass collection tube, the genomic mouse DNA is eluted with 7 mL of Buffer QF (1.25 M NaCl; 50 mM Tris-Cl, pH 8.5; and 15% isopropanol) that has been heated to 50 °C. To the eluate add 5 mL of isopropanol.

370. Why is the DNA in step 2, above, eluted?

a. the molarity of the NaCl is increased.
b. the continuous presence of the 15% isopropanol saturates the column, which releases the DNA
c. the pH is raised to 8.5
d. the buffer type is changed from MOPS to Tris
e. both *a* and *c*

Step 3. The collection tube from step 2, above, is centrifuged for 30 min at 5,000 × g, the supernatant is poured off, and the DNA pellet is washed with 2 mL ice-cold 70% ethanol. The collection tube is again centrifuged at 5,000 × g for 5 min, the wash is poured off, and the tube is air-dried by inverting it over a dry Kimwipe. The DNA pellet is gently re-suspended in TNE Buffer, first overnight at 40 °C and then for an additional 1–7 d at room temperature. (NOTE: Longer resuspension times will result in more of the DNA from the pellet going into solution. The DNA pellet must be resuspended gently in order to recover high-molecular-weight genomic DNA.)

Step 4. A 45-mm-diameter Millipore membrane (VSWP 04500, pore size 0.025 μM, Type VS) is floated on 200 mL of 0.1 X TE buffer in a

250 mL beaker. After wetting for about 5 min, 500 μL of the DNA in TE buffer from step 3, above, is pipetted onto the center of the Millipore membrane, the beaker is covered with half of a Petri dish, and dialysis is allowed to proceed overnight (at least 5–6 h). After this drop dialysis, as much of the liquid remaining on the Millipore filter as possible is pipetted into a plastic centrifuge tube on ice.

371. What is the purpose of the dialysis in step 4?

a. to remove excess salts, amino acids, and other small molecules from the isolated DNA
b. to dehydrate the DNA that is primarily in an ethanolic suspension
c. to re-associate RNA with the genomic DNA
d. to re-associate the necessary histones with the genomic DNA
e. to re-form double-stranded DNA from the single strands that remain after column cleanup

The purity of the DNA is estimated by scanning from 220 to 320 nM in a UV spectrophotomer. The ratio of the readings at 260 nM and 280 nM is found to be 1.8.

372. What does the above ratio indicate?

a. that the DNA is contaminated with RNA
b. that the DNA is contaminated with protein
c. that the DNA is contaminated with protein and RNA
d. that the DNA has an acceptable purity
e. that the DNA is still contaminated with buffer salts and the dialysis needs to be repeated

373. If the ratio of the readings at 260 nM and 280 nM were 1.5, what would one conclude?

a. that the DNA is contaminated with RNA
b. that the DNA is contaminated with protein, lipid, and/or other cellular contaminants
c. that the DNA is contaminated with protein and RNA
d. that the DNA has an acceptable purity
e. that the DNA is still contaminated with buffer salts and the dialysis needs to be repeated

374. If the ratio of the readings at 260 nM and 280 nM were 2.0, what would one conclude?

a. that the DNA is contaminated with RNA
b. that the DNA is contaminated with protein
c. that the DNA is contaminated with protein and RNA
d. that the DNA has an acceptable purity
e. that the DNA is still contaminated with buffer salts and the dialysis needs to be repeated

Step 5. Set up the following six digests (A to F) of the mouse DNA:

A

2 μL mouse DNA (1 μg)
5 μL 10X NEB Buffer 1
40 μL $H_2O$
3 μL *Alu*I enzyme

Incubate at 37 °C overnight

375. Which of the following is/are *true* about the *Alu*I enzyme?

   a. It is a "four-cutter."
   b. It cuts at the phosphodiester linkage between G and C at the GC junction in 5′-AGCT-3′/3′-TCGA-5′ pairs.
   c. It is not sensitive to *dam*, *dcm*, or mammalian CpG methylation.
   d. All of the above are true.
   e. Only *a* and *b* are true.

B

2 μL mouse DNA (1 μg)
5 μL 10X *Eco*RI Buffer
40 μL $H_2O$
3 μL *Eco*RI enzyme

Incubate at 37 °C overnight.

C

2 μL mouse DNA (1 μg)
5 μL 10X NEB Buffer 2
40 μL $H_2O$
3 μL *Sfi*I enzyme

Incubate at 50 °C overnight.

376. Which of the following is/are *true* about the *Sfi*I enzyme?

   a. It may have cleavage blocked by methylation at a specific CpG site located within the recognition sequence.
   b. It cuts at the first and second "N" of the 5′-GGCCNNNNNGGCC-3′/3′-CCGGNNNNNCCGG-5′.
   c. It cannot be heat inactivated at 65 °C for 20 min.
   d. It may have cleavage blocked by dcm methylase activity.
   e. All of the above are true.

D

2 μL mouse DNA (1 μg)
5 μL 10X NEB Buffer 1
40 μL $H_2O$
3 μL *Hpa*II enzyme

Incubate at 37 °C overnight

377. Which of the following is *true* about the *Hpa* II enzyme?

a. It is a "four cutter."
b. It is a "six cutter" like *Hpa* I.
c. It cuts at T-A in 5′-GTTAAC-3′/3′-CAATTG-5′.
d. It leaves blunt ends (i.e., no overhang).
e. Like HpaI, it cannot be heat inactivated at 65 °C for 20 min.

E
2 μL mouse DNA (1 μg)
5 μL 10X NEB Buffer 2
40 μL $H_2O$
3 μL *Msp*I enzyme

Incubate at 37 °C overnight.

378. Which of the following is/are *true* about the enzyme *Msp*I?

a. It is an isoschizomer of *Hpa*II.
b. It is not sensitive to *dam*, *dcm*, or mammalian CpG methylation.
c. It is a "four cutter."
d. It leaves a CGG overhang at the cleavage site after cutting.
e. All of the above are true.

F
2 μL mouse DNA (1 μg)
5 μL 10X NEB Buffer 2
43 μL $H_2O$

Incubate at 37 °C overnight.

379. True or false: The mixture without enzyme is put in to provide a negative control.

Step 6. To each digest tube from step 5 (A–F) above, add 5 μL of 3.0 M sodium acetate (pH 5.2) and 100 μL ice-cold 100% ethanol and mix. Centrifuge all of the mouse digest tubes. Without disturbing the pellet, remove the supernatant from each digest tube, add 500 μL 70% ethanol to each tube, invert each tube several times, and centrifuge. Again, remove the supernatant from each tube. To each pellet in each tube add 12 μL of 1X TE buffer plus 3 μL of 5X "Blue Juice." Allow the tubes to sit overnight at 4 °C.

Step 7. Perform a PCR using the following mixture. The purpose of this PCR is to amplify the single copy *Ttr* gene from total genomic mouse DNA.

30.5 μL $H_2O$
8 μL dNTP mixture
5 μL 10X PCR reaction mixture
3 μL $MgCl_2$
1 μL *Ttr* Primer Forward 1 (10 pmoles/μL)
1 μL *Ttr* Primer Reverse 1 (10 pmoles/μL)
1 μL template DNA (uncut mouse DNA isolated above)

Use the following PCR conditions:

An initial denaturing step at 94 °C for 12 min

The following steps are cycled 35 times:

94 °C for 30 sec to denature dsDNA.
58 °C for 30 sec to enable primers to anneal to template DNA.
72 °C for 1 min to synthesize new DNA.

A final extension/annealing step at 72 °C to insure that all extensions are complete.

Hold at 4°C until the instrument is turned off.

Step 8. Pour another 1% agarose gel and allow it to solidify for at least 20 min before adding the TAE running buffer. This will be the Southern Blot gel. Retrieve 5 μL of λ*Bst*EII marker, heat at 65 °C for 5 min, and place on ice. Load the 1% agarose gel in the following order:

10 μL *Alu*I-cut mouse DNA in "Blue Juice."
10 μL *Eco*RI-cut mouse DNA in "Blue Juice."
10 μL *Sfi*I-cut mouse DNA in "Blue Juice."
10 μL *Hpa*II-cut mouse DNA in "Blue Juice."
10 μL *Msp*I-cut mouse DNA in "Blue Juice."
10 μL uncut mouse DNA + 5 μL λ*Bst*EII marker in "Blue Juice."

Electrophorese at 50 VDC for 1 h.

Retrieve the *Ttr* PCR product from step 7, above. Dilute 5 μL with 95 μL TE buffer. To a separate tube add 1 μL of the undiluted PCR product and 7 μL of the diluted PCR product, 3 μL of TE Buffer, and 2 μL of 5X "Blue Juice." Load this sample in the gel in lane 8.

Also retrieve 1 μL of the product of a reverse transcriptase PCR of mouse 1.4 kb *Eco*RI-cut DNA and mix with 7 μL of 1X TE buffer and 2 μL 5X "Blue Juice." Load this sample in the gel in lane 9.

Load 10 μL of a 100 bp ladder in "Blue Juice" in the gel in lane 10 and, in separate lanes, the three products from the above steps onto the open lanes of the gel. Electrophorese for 90 min.

Step 9. After photographing, save the gel and then add enough 0.25 N HCl to cover the gel and allow to sit for 15 min. Pour off the HCl and rinse once with $H_2O$.

380. Why is the HCl added?

a. to neutralize the running buffer
b. to neutralize $OH^-$ produced during the electrophoretic process
c. to depyrimidate the DNA
d. to depurinate the DNA
e. to denature the agarose to make the DNA more accessible

Add enough 0.5 N NaOH to cover the gel and incubate for 20 min at room temperature. Pour off the NaOH and rinse once with $H_2O$. Neutralize the gel with 1.0 M Tris-HCl (pH 7.5) in 1.5 M NaCl for 20 min.

381. Why is the NaOH added?

a. to denature the DNA in the gel
b. only to neutralize the HCl that was added above
c. depyrimidate the DNA
d. depurinate the DNA
e. denature the agarose to make the DNA more accessible

Step 10. Wet a nylon membrane first with $H_2O$ for 5 min and then with 10X SSC (0.3 M trisodium citrate, 3 M NaCl, pH 7.0) for 15 min. Using Whatman 3MM paper, create a filter paper base with three sheets of paper and a wick with one larger sheet of the same filter paper. Saturate the base and wick with 10X SSC. Place the gel from step 9, above, face down on the filter paper wick. Place the wetted nylon membrane from this step on the gel. Place another piece of Whatman paper that has been saturated with 10X SSC on the nylon membrane and then place two sheets of dry filter paper on the wet piece of filter paper. Cover the filter paper with a stack of dry paper towels that have been cut to the same size as the gel and Whatman papers. Place a rigid support on top of the paper towel stack and place a 200-g to 500-g weight on the rigid support. Allow the transfer to proceed for a minimum of 4–6 h.

382. True or false: In assembling the Southern blot transfer apparatus, it is critical that all bubbles be removed.

Step 11. After transfer is complete, rinse the membrane in 6X SSC for 2–5 min.

Step 12. Check the gel for fluorescence.

383. True or false: There should be intense fluorescence remaining in the gel, indicating that the DNA has transferred and the small molecule ethidium bromide has stayed behind in the agarose (successful blot).

Step 13. Place the rinsed membrane from step 11, above, on a clean sheet of filter paper and allow to air dry.

Step 14. Subject the membrane to UV light and then bake at 80 °C for 1 h in an oven. Cut the membrane between lanes 7 and 8.

384. True or false: The baking step requires a vacuum oven because the nylon membrane can autoignite like nitrocellulose.

Step 15. To a tube containing 210 ng of mouse *Ttr*-specific oligonucleotide probe, add the following components:

24.5 μL $H_2O$
2.5 μL Fluorescein-11-dUTP
4.0 μL 10X Cacodylate Buffer
4.0 μL Terminal Transferase Enzyme

Mix the components and allow to incubate for 90 min at 37 °C.

385. What is the purpose of the terminal transferase?

a. to add several fluorescein-labeled uridine nucleotides to the 3′-hydroxyl terminus of the *Ttr*-specific oligonucleotide
b. to remove the 3′-phosphate group
c. to add a fluorescein-labeled uridine to the 5′-hydroxyl terminus of the *Ttr*-specific oligonucleotide
d. to remove the 5′-phosphate group
e. to accomplish *a* and *b* contemporaneously

Step 16. Place the *Ttr*/PCR Southern part of the membrane (containing lanes 8–10) from step 14, above, into a plastic hybridization bag and add 4 mL of Oligo Hyb Solution (5X SSC, 0.1% Hyb buffer component, 0.02% sodium dodecyl sulfate or SDS, and a 20-fold dilution of liquid block). Prehybridize for 1–2 h at 55 °C.

Step 17. Place the genomic part of the membrane (containing the DNA from lanes 1–7) in another, separate bag and add 4 mL of Probe-Amp Hybrization Buffer (5X SSC, 5.0% Dextran Sulfate, 0.5% Carnation® Nonfat Dry Milk, 0.1% SDS, and 100 μg/mL denatured salmon sperm DNA). Incubate at 70 °C for 1–6 h.

386. DNA that blocks "sticky" or non-specific binding sites is provided by which of the following?

a. dextran sulfate
b. Carnation® Nonfat Dry Milk
c. protein left over from step 17, above
d. denatured salmon sperm DNA
e. the Probe-Amp Hybridization Buffer

Step 18. Cut the bag from step 16, above, and add 20 μL of *Ttr* oligonucleotide probe from step 15, above. Reseal the bag and hybridize at 55 °C for 2–17 h.

387. True or false: In steps 16–18 it is critical that at least a few air bubbles be left, to provide proper mixing when the bag is rocked.

Step 19. Mix 4 μL of a mouse *Rvt* (reverse transcriptase), fluorescein-labeled probe from the LINE-1 repeat (separately prepared) with 46 μL of Hyb buffer, heat at 95 °C for 5 min, centrifuge, and put on ice immediately. Again, cut the bag from step 17, above, and place all 46 μL of the fluorescein-labeled probe into the bag. Reseal and hybridize at 70 °C overnight with shaking.

Step 20. Remove the filters (lanes 8–10) from the bag with the *Ttr* oligonucleotide probe and wash with 5X SSC, 0.1% SDS, at room temperature with shaking for 5 min. Repeat the wash in 5X SSC one more time. Repeat the wash in 1X SSC, 0.1% SDS, for 30 min at 55 °C with shaking.

Step 21. Wash the filter once in 50 mL of ECL Buffer 1 (0.15 M NaCl, 0.1 M Tris base, adjusted to pH 7.5 with concentrated HCl) at room temperature. Discard this wash solution and incubate for at least 30 min at room temperature with 50 mL of blocking solution. Discard the blocking solution and wash again in ECL Buffer 1.

Step 22. Dilute anti-fluorescein antibody/horseradish peroxidase conjugate 1000-fold in ECL Buffer 2 (0.4 M NaCl, 0.1 M Tris base, adjusted to pH 7.5 with concentrated HCl) containing 0.5% bovine serum albumin. Incubate the membranes for 30 min in 50 mL of the diluted antibody conjugate.

388. The fluorescein label on the oligonucleotide probe does what?

a. It provides a fluorescent label for the fluorescent detection of the mouse *Ttr* gene.
b. It blocks the 3′ end of the oligonucleotide from the addition of further nucleotides.
c. It blocks the 5′ end of the oligonucleotide from the addition of further nucleotides.
d. It provides an antigenic site for the anti-fluorescein antibody/horseradish peroxidase conjugate that will be used for detection.
e. It blocks any further hybridization by unwanted, non-specific nucleotides.

Step 23. Rinse the membranes a total of four times in separate 50 mL volumes of ECL Buffer 2.

Step 24. Drain off the ECL Buffer 2 and add about 2 mL of a mixture of detection solution 1 and detection solution 2 to the surface of each membrane. Expose x-ray film to each membrane.

389. The presence of the mouse *Ttr* PCR products, if present, will be detected using which of the following?

a. chemiluminescence generated by the horseradish peroxidase acting on the components in the two detection solutions
b. the natural fluorescence of the fluorescein
c. induced chemiluminesce produced by fluorescein
d. the natural fluorescence of the uridine added in step 21
e. the induced chemiluminescence of the uridine added in step 21

Step 25. Cut open the hybridization bag containing the genomic part of the membrane (lanes 1–7) and wash for 15 min at 70 °C in 1X SSC, 0.1% SDS. Replace the wash with 0.5X SSC, 0.1% SDS, and wash for 20 min at 70 °C. Rinse the membrane briefly in ECL Buffer 1 and then incubate for at least 1 h in blocking solution at room temperature. Pour off the solution, rinse briefly in ECL Buffer 1, and pour off the solution.

Step 26. Incubate the membrane in 50 mL of antibody solution for 0.5–1 h at room temperature with agitation. Pour off the antibody solution and wash twice in 0.1% Tween in ECL Buffer 1 at room temperature followed by two additional washes for 5 min each.

Step 27. Mix an equal volume of ECL detection solutions 1 and 2 to give sufficient volume to cover the membrane.

Step 28. Remove the membrane from the last wash solution and drain off the excess. Cover the membrane with the detection solution mixture from step 26, above. Incubate for 1 min.

Step 29. Drain off the excess and place in an acetate development cassette. With the DNA side up, place a sheet of x-ray film on the top of the blot. Expose the film for 3–10 min, develop, and read the lumigram obtained.

390. Which of the following is/are *true* about a 2250G>A mutation of the *ATM* gene (AAG becomes AAA at the end of exon 16)?

a. It results in no change at codon 709 (lysine is still produced).
b. It is considered a silent mutation if only the genomic analysis is considered.
c. It is a "crosswalk" transition.
d. all of the above
e. only *a* and *b*

391. Referring to question 390, above, protein truncation testing (PTT) reveals a failure to produce any ATM protein in a patient with the 2250G>A mutation. Why?

a. because the mutation results in a stop codon
b. because the mutation results in a frameshift that produces a protein totally unrelated to ATM
c. because the mutation perturbs the donor splice site of exon 16, causing the acceptor site to skip to the end of exon 15 and deleting the entire 126 bp of exon 16
d. because the mutation inserts a phenylalanine where lysine was, causing a protein to fold completely differently than the normal ATM protein
e. because the mutation causes the repeat of exon 16 and produces a protein that is too long to be functional and does not manifest itself on traditional western blots

392. Given a segment of dsDNA that is 14,514 bases in length, that has a %G-C content of 45%, and that is in a solution of 2.0 M $Na^+$ and 10% formamide, calculate the segment's melting temperature ($T_M$) in °C.

________________________________________

393. Peptidyl transferase activity in the ribosome is located where?

a. in the small subunit exclusively
b. It is shared between the large and small subunits.
c. It is part of the proteins of the large subunit (it is an enzyme, after all).
d. It is a ribozyme activity that is located in the large subunit.

394. Which of the following factors has a 3-D shape and size similar to the RF3/RF2 complex?

a. EF-Tu + tRNA
b. EF-G
c. EF-Ts
d. both *a* and *b*

395. The *PTEN* gene (10g23) is a tumor-suppressor gene that frequently is mutated in which of the following?

a. glioblastoma
b. prostate cancers
c. breast cancers
d. all of the above
e. Only *b* and *c*

396. You have been given a 1-mm cube of fresh tissue from which to extract DNA. After extracting the entire specimen, you resuspend the DNA in 50 μL of buffer. Using 10 μL in 990 μL buffer for spectrophotometry, the $A_{260} = 0.5$ and $A_{280} = 0.28$. What is the concentration of DNA?

a. 2.5 μg/μL
b. 5 μg/μL
c. 25 μg/μL
d. none of the above

397. In a chromosome walking experiment, a library of 96 separate clones is probed with clone A1; positive results are observed with clones E7 and F6, as well as with A1. Interpret these results.

a. Clones A1, E7, and F6 are identical in all respects.
b. Clone A1 overlaps with clones E7 and F6.
c. Clones E7 and F6 are identical, but different from clone A1.
d. No interpretation is possible from these data.

398. Using standard dsDNA terminology in which the top strand is 5′→3′ and the bottom strand is 3′→5′, which strand has the same sequence as the mRNA generated from it, assuming that it is the bottom strand that is read and that it is read 5′→3′?

a. the top strand
b. the bottom strand
c. neither strand
d. both strands

399. Different types of RNA are critical cellular components. Which of the following is *not* a type of RNA found in cells?

a. mRNA
b. rRNA
c. tRNA
d. cRNA

400. The prokaryotic ribosome is a 2-subunit particle described as 70S. Identify the subunits.

a. 30S and 40S
b. 30S and 50S
c. 40S and 60S
d. 16S and 20S

401. The random primer reaction is typically carried out at room temperature, rather than at 37 °C. Why?

a. DNase I is more active at 15 °C, and so more nicks are formed.
b. The hexamer primers will not anneal at 37 °C, but will at room temperature.
c. To slow down the activity of the DNase I, preventing the formation of too many nicks.
d. This is a false statement; random priming is performed at 42 °C.

402. What is the key principle behind the strand displacement amplification reaction?

a. Incorporation of thio-dCTP prevents newly synthesized strands from being restriction digested.
b. The 5′-3′ exonuclease activity of DNA polymerase I digests the 3′ DNA as the nick is extended from the restriction enzyme digestion site.
c. Primers anneal at lower temperatures. These primers are displaced as DNA synthesis occurs.
d. DNA melts in different regions of the strand, depending upon the AT and GC content.

403. You wish to prepare a riboprobe. Which labeled nucleotide can you use?

1. thio-dCTP
2. $\alpha$-[$^{32}$P]-UTP
3. digoxigenin-11-dUTP
4. biotin-16-UTP

a. 1, 2, and 3 are correct.
b. 1 and 3 only are correct.
c. 2 and 4 only are correct.
d. 4 only is correct.

404. *Taq* polymerase can be used to prepare a strand-specific probe. How?

a. through the use of one primer and incorporation of $\alpha$-[$^{32}$P]-dUTP
b. through the use of both fluorescently labeled upstream and downstream primers
c. through the incorporation of biotin-16-UTP and a single primer (primer extension reaction)
d. It can't. This is a false statement.

405. RT-PCR consists of several steps. What is the first one?

a. annealing of an appropriate strand-specific primer to the gene of interest
b. melting of DNA
c. reverse-transcription of the RNA to form cDNA
d. extension of a biotin labeled primer

406. The Hybrid Capture Assay relies upon which of the following principles?

a. capture of an RNA:DNA duplex molecule by a bound antibody
b. annealing of target DNA to a bound capture probe
c. extension of a bumper primer to displace a hybrid DNA molecule
d. relief of fluorescence energy transfer quenching as extension proceeds

407. In the *gapped* LCR procedure, which two additional components need to be included in the reaction compared to standard LCR?

a. NTPs
b. Taq polymerase
c. Klenow fragment
d. dNTPs
e. both *a* and *b*
f. both *a* and *c*
g. both *b* and *d*
h. both *c* and *d*

408. Which of the following amplification reactions is/are isothermal?

1. SSCP
2. TMA
3. LCR
4. SDA

a. 1, 2, and 3 are correct.
b. 1 and 3 only are correct.
c. 2 and 4 only are correct.
d. 4 only is correct.

409. Which of the following labeling reactions *cannot* yield a strand-specific probe from a double-stranded template?

1. nick translation
2. 3′ end labeling with TdT
3. random primer labeling
4. riboprobes

a. 1, 2, and 3 are correct.
b. 1 and 3 only are correct.
c. 2 and 4 only are correct.
d. 4 only is correct.

410. Which of the following is/are not needed in the SDA procedure?

a. strand-specific primers
b. α-[$^{32}$P]-dCTP
c. bumper primers
d. thio-dCTP

411. In the clone fingerprinting technique, physical information is obtained about each cloned fragment; similarities indicate possible overlaps. Possible physical information can include which of the following?

a. restriction fragment patterns
b. repetitive DNA fingerprints
c. STS content mapping
d. all of the above

412. The C3435T mutation in the *MDR1* gene is associated with which of the following?

a. reduced intestinal P-glycoprotein and higher plasma concentrations of the P-glycoprotein substrate digoxin
b. increased intestinal P-glycoprotein and reduced plasma concentrations of the P-glycoprotein substrate digoxin
c. increased intestinal P-glycoprotein and increased plasma concentrations of the P-glycoprotein substrate digoxin
d. reduced intestinal P-glycoprotein and reduced plasma concentrations of the P-glycoprotein substrate digoxin
e. reduced intestinal P-glycoprotein but no change in plasma levels of the P-glycoprotein substrate digoxin due to increased binding of the mutant protein

413. Given two markers that are 5 cM apart, calculate the probability of a true double crossover between the two markers.

____________%

414. Which of the following is *not* a method for transferring DNA from an agarose gel to a blotting membrane?

a. vacuum transfer
b. siphon transfer
c. electroblotting
d. capillary transfer

415. You have run an agarose gel in 1X TBE, pH 8.3. In order to transfer the DNA to a membrane, what must be done next?

a. Immediately set up the transfer with 5X SSC, pH 7.0.
b. Soak the gel in 0.2 M HCl then 0.5 M NaOH, followed immediately by transfer in 5X SSC, pH 7.0.
c. Soak the gel in 1 M NaCl, followed immediately by transfer in 5X SSC, pH 7.0.
d. Immediately set up the transfer with 0.4M NaOH, pH 13.6.

416. Functional polymorphism at the *ADH1B* (alcohol dehydrogenase) gene will lead to which of the following?

a. ADH3B*1, ADH3B*2, and ADH3B*3
b. ADH1C*1 and ADH1C*2
c. ADH5, ADH6, and ADH7
d. ADH1B*1, ADH1B*2, and ADH1B*3
e. None of the above, because *ADH1B* does not demonstrate any polymorphism.

417. You have transferred DNA to an uncharged nylon membrane using 0.4M NaOH. You immediately proceed with the hybridization portion of the experiment. Predict the results.

a. You will detect your band of interest following development of the probe.
b. The DNA will wash off during the hybridization process because it is not fixed in place.
c. If you use a riboprobe, you will detect nothing because RNA will not hybridize to DNA.
d. If you use a riboprobe, residual NaOH will degrade the probe and so you will detect nothing.

418. *Increasing* which of the following components will *decrease* the melting temperature of a DNA duplex?

a. salt concentration
b. formamide concentration
c. GC content of the probe
d. length of the probe
e. both *a* and *b*

419. You have prepared a nick-translated probe, but forgot to boil it prior to adding it to your hybridization solution. Predict the results.

a. The probe will hybridize normally, as long as the membrane was properly fixed.
b. The probe will hybridize normally, as long as the formamide concentration is kept low.
c. The probe will hybridize poorly, if at all, since it is double stranded already.
d. The probe will hybridize normally as long as the salt concentration is kept high.

420. The DPD (dihydropyrimidine dehydrogenase, which inactivates 5-fluorouracil) genotype is ______________________________?

a. autosomal recessive
b. X-linked dominant
c. mitochondrially inherited
d. X-linked recessive
e. autosomal dominant

421. The dystrophin (*DMD*) gene is often mutated by the deletion of one or more exons. Which of the following is a convenient way to detect these deletions?

a. RT-PCR
b. SDA
c. multiplex PCR
d. the protein truncation test

422. A mutation of a DNA sequence in the genome that changes GAATTC to GAATTT can be termed what?

a. a restriction fragment length polymorphism
b. a silent mutation
c. a nonsense mutation
d. a frameshift mutation

423. Considering the TΨC arm of tRNA, which of the following is a false statement?

a. There are six base pairs in the stem.
b. The base pseudouridine is found.
c. The loop contains the invariant sequence TΨC.
d. It is located 3′ to the D arm.

424. The wild-type upstream primer in a PCR method is designed such that the 3′ base is a G, corresponding to a C in the wild-type allele; the mutant allele has a T at this position. The mutant upstream primer is identical in sequence to this primer except that the 3′ base is an A. The downstream primer sequence is common to both wild-type and mutant alleles. Two separate PCR assays are run, one with wild-type upstream primer and the other with mutant upstream primer. Predict the results if the patient was heterozygous for the mutation.

a. expected band in the wild-type reaction only
b. expected band in the mutant reaction only
c. expected bands in both reactions
d. expected band in neither reaction

425. Trinucleotide expansions are mutations commonly associated with which type of disease?

a. coagulation disorders
b. neurological disorders
c. lipoprotein excesses
d. cystic fibrosis

426. You wish to screen a cystic fibrosis patient for any of the several known mutations that typically cause this disease. Which procedure could you conceivably use?

a. forward ASO hybridization
b. reverse ASO hybridization
c. DGGE
d. SSCP

427. Which of the following mutation detection assays is performed using denaturing gels?

a. heteroduplex analysis
b. SSCP
c. DGGE
d. ASO hybridization

428. Predict the effect of increasing the salt concentration on the $T_m$ of a labeled oligonucleotide probe.

a. It will be increased.
b. It will be decreased.
c. There will be no effect.
d. It will increase, but only if formamide is included in the reaction.

429. Which of the following agents is/are typically used to denature DNA?

1. formamide
2. heat
3. urea
4. NaCl

a. 1, 2, and 3 are correct.
b. 1 and 3 only are correct.
c. 2 and 4 only are correct.
d. 4 only is correct

**430–438. Match the following.**

| | |
|---|---|
| 430. _____ Transcription | a. polymerase encased inside a wax bead |
| 431. _____ Ethidium bromide | b. hapten used to label DNA or RNA |
| 432. _____ Intron | c. creates an AP site |
| 433. _____ Hybridization | d. untranslated DNA sequence |
| 434. _____ MOPS buffer, pH 7.0 | e. annealing of probe to template |
| 435. _____ B-form | f. RNA electrophoresis |
| 436. _____ Uracil-N-glycosylase | g. most common form of DNA |
| 437. _____ Hot-start PCR | h. fluorescent detection of DNA |
| 438. _____ Digoxigenin | i. synthesis of RNA using a DNA template |

439. A prokaryotic messenger RNA contains several characteristic segments that are necessary for its function. Which of the listings below is in the correct order?

a. promoter, Shine-Dalgarno sequence, start codon, stop codon, stem-loop
b. promoter, transcriptional start point, leader, ORFs preceded by a Shine-Dalgarno sequence and separated by intercistronic sequences, trailer, stem-loop
c. transcriptional start point, ORFs preceded by a Shine-Dalgarno sequence and separated by intercistronic sequences, trailer, stem-loop, and stop codon
d. transcriptional start point, ORFs preceded by a Shine-Dalgarno sequence and separated by intercistronic sequences, trailer. Each ORF has its own start and stop codon
e. transcriptional start point, leader, one Pribnow Box sequence followed by ORFs separated by intercistronic sequences, and trailer

440. Which of the following is *not* characteristic of eukaryotic mRNA?

a. the presence of an inverted "G" cap at the 5′ end
b. the presence of up to 200 A residues at the 3′ end
c. splicing to remove intron sequences and to join the exon sequences
d. typically monocistronic (codes for only a single protein)
e. processed in the nucleus and then transported to the cytoplasm

441. What is a ribosome?

a. an rDNA-containing organelle in which rRNA synthesis and mRNA translation occur
b. a macromolecular, two-subunit complex containing more than 50 different proteins and several rRNAs
c. a nuclear protcin complex that facilitates protein synthesis
d. a structure, consisting of three (prokaryotes) or four (eukaryotes) RNAs, that facilitates protein synthesis
e. an RNA structure that guides the template selectivity of RNA polymerases I, II, and III in eukaryotic cells

442. In eukaryotes, what is the function of eIF-3 in translation?

a. It facilitates the formation of the subunit initiation complex (eIF-3, GTP, Met-tRNA, 40S).
b. It facilitates the binding of the subunit initiation complex (ternary complex, 40S) to the 5′ end of the mRNA.
c. If bound to the 40S subunit, it prevents the association of the 40S and 60S subunits.
d. It binds to the cap at the 5′ end of the mRNA in order to unwind secondary structure.
e. It activates the ribosomal GTPase, which enables subunit association upon GTP hydrolysis and release of eIF-2.

443. Which of the following statements about the ribosomal peptidyl transferase activity is correct?

a. The peptidyl transferase activity resides in the large (50S or 60S) ribosomal subunit.
b. It facilitates transfer of the C-terminus of the peptide chain from the peptidyl-tRNA to the N-terminus of the aminoacyl-tRNA in the A-site.
c. It facilitates transfer of the N-terminus of the aminoacyl-tRNA from the A-site to the C-terminus of the peptidyl-tRNA in the P-site by deacylation of aminoacyl-tRNA.
d. It hydrolyzes GTP in order to allow for translocation of the ribosome.
e. It deacylates peptidyl-tRNA.

444. Termination of ribosomal peptide synthesis (translation) is achieved by which of the following?

a. in-frame occurrence of a codon that codes for the C-terminal amino acid
b. in-frame occurrence of a codon for which no cognate peptidyl tranferase exists
c. a too-low concentration or the lack of a particular aminoacyl-tRNA
d. GTP-dependent action of release factors (RFs) that prevent mismatch binding between termination codons and aminoacyl-tRNAs in the A-site
e. the activity of terminal peptidyl transferase, a protein that deacylates peptidyl-tRNA by adding a lysine or arginine residue to the C-terminus of the nascent polypeptide

445. Which of the following rRNA molecules are found in eukaryotic ribosomes?

a. 5.8S
b. 5S
c. 16S
d. 23S
e. 28S
f. both *a* and *b*
g. *a*, *b*, and *d*
h. *a*, *b*, and *e*
i. both *c* and *d*

446. Which base pairs are found in double-stranded DNA?

a. A • U
b. G • T
c. C • G
d. T • A
e. C • A
f. both *a* and *b*
g. both *c* and *d*
h. answers *a*, *b*, and *e*

447. Which of the following is true about the replication of DNA?

a. It involves the synthesis of two daughter strands that associate together in a completely new double helix (conservative replication).
b. It follows the principle that a new daughter strand is assembled on each parental strand (semi-conservative replication).
c. It is dependent on the species-specific genetic code.
d. It is the most common source for base mispairing.
e. It is a process that describes gene expression.

448. Two experiments were crucial for the identification of DNA as the genetic material: a study of bacterial (*Streptococcus*) virulence in mice and the infection of *E. coli* bacteria by bacteriophage T2. What was the major argument in both cases?

a. DNA was re-isolated from infected organisms as the "causal agent" of disease.
b. Mutated DNA caused a failure of virulence.
c. DNA—but not protein—taken up from an external source by an organism changed its genetic potential.
d. DNA was not transferable between organisms and hence must be a very conservative molecule.
e. DNA from prokaryotes, eukaryotes, and viruses could be mixed and substituted for one another.

449. In eukaryotic cells, DNA replication occurs in the ______________ of the cell cycle.

a. G1 phase
b. S phase
c. G2 phase
d. M phase
e. G0 phase

450. DNA replication proceeds in the ___________ direction, similar to RNA transcription.

a. $5' \rightarrow 3'$
b. $3' \rightarrow 5'$
c. $N \rightarrow C$
d. opposite

451. Prokaryotic replicons are typically what?

a. linear, with a single origin of replication
b. linear, with multiple origins of replication
c. circular, with a single origin of replication
d. circular, with multiple origins of replication

452. DNA replication has been studied in temperature-sensitive mutants in order to identify the genes and the functions of the proteins involved in the process. Some mutants are termed *quick-stop* mutants because they cease replication immediately upon a rise in temperature. This indicated that the mutated protein is likely involved in replication ___________________.

a. initiation
b. elongation
c. termination
d. concatenation

453. *E. coli* has three separate and distinct DNA polymerase enzymes. Which one is *primarily* involved in the process of DNA repair?

a. DNA polymerase I
b. DNA polymerase II
c. DNA polymerase III
d. DNA polymerase β

454. DNA polymerase I of *E. coli* has three separate activities (impressive for a single protein, don't you think?!). These are which of the following?

a. $3' \rightarrow 5'$ synthesis, $5' \rightarrow 3'$ exonuclease, $3' \rightarrow 5'$ exonuclease
b. $3' \rightarrow 5'$ synthesis, $5' \rightarrow 3'$ exonuclease, $3' \rightarrow 5'$ exonuclease
c. $3' \rightarrow 5'$ synthesis, $5' \rightarrow 3'$ exonuclease, $3' \rightarrow 5'$ endonuclease
d. $5' \rightarrow 3'$ synthesis, $5' \rightarrow 3'$ exonuclease, $3' \rightarrow 5'$ exonuclease

455. Which of the following subunits of *E. coli* DNA polymerase III contains the synthetic activity?

a. the α subunit
b. the γ subunit
c. the ε subunit
d. the β subunit
e. none of the above

456. DNA polymerases have been isolated and studied from a variety of organisms, from viruses to bacteria to chickens, frogs, and mammals. What is one feature that they all have in common?

a. They are all multisubunit proteins.
b. They all require a primer.
c. They all synthesize DNA from an RNA template.
d. They all possess $5' \rightarrow 3'$ exonuclease activity.

457. DNA replication is termed _________________ replication.

a. semidiscontinuous
b. conservative
c. discontinuous
d. high fidelity

458. In *E. coli* DNA replication, which protein lays down the RNA primer?

a. DnaB
b. DnaE
c. DnaQ
d. DnaG

459. In eukaryotes, homologous recombination typically occurs when?

a. during S phase
b. during mitosis
c. during meiosis
d. during excision repair

460. Which prokaryotic enzyme system is responsible for branch migration during recombination in *E. coli*?

a. RecA
b. RecBCD
c. endonuclease HO
d. RuvAB

461. Which of the following are necessary for λ phage DNA to integrate into the *E. coli* genome?

a. *Xis* gene product
b. IHF
c. *Int* gene product
d. Only *b* and *c* are correct.
e. Answers *a*, *b*, and *c* are all correct.

462. Which of the following are components of the mouse mammary tumor virus (MMTV) particle?

a. two identical copies of dsRNA
b. integrase
c. reverse transcriptase
d. all of the above
e. Only *a* and *c* are correct.

463. Which of the following genes is *not* found in retroviruses?

a. *Gag*
b. *Pol*
c. *Env*
d. *Onc*

464. Which of the following is used as a primer for reverse transcriptase?

a. integrase protein covalently linked to a G nucleotide
b. an uncharged host tRNA
c. a hairpin structure at the 3′ end of the retroviral genome
d. a single-stranded break

465. Where is the promoter for the *env* gene of the MMTV provirus?

a. in the adjacent host DNA
b. in the upstream LTR
c. in the downstream LTR
d. in the middle of the *pol* gene

466. Which of the following is *not* produced by an α type haploid yeast?

a. **a** factor
b. α factor
c. STE3 protein
d. HO endonuclease

467. Where is α genetic information found?

a. in the *HML* and *MAT* loci
b. in the *HML* and *HMR* loci
c. in the *MAT* and *HMR* loci
d. in the *HML*, *MAT*, and *HMR* loci

468. Which of the following statements about *HO* gene expression is *false*?

a. The *HO* gene is expressed in both haploids and diploids.
b. The *HO* gene is transcribed in mother cells, but not in daughter cells.
c. The *HO* gene is expressed only at the end of G1.
d. HO cleaves the MAT locus just 3′ of the Y region.

469. What features must every eukaryotic chromosome contain?

a. centromere
b. telomere
c. origin of replication
d. all of the above

470. What is the approximate size of the diploid human genome?

a. 3 billion base pairs
b. 6 billion base pairs
c. 9 billion base pairs
d. 12 billion base pairs

471. Which of the following is *not* associated with DNA that attaches to the nuclear matrix?

a. topoisomerase II recognition site
b. *Cis* acting transcriptional regulation sites
c. 70% A•T rich
d. the nucleoid

472. "Chromatin that is not transcribed, is replicated late in S-phase, and is typically found at the centromere and elsewhere"—this definition describes which of the following?

a. heterochromatin
b. euchromatin
c. the nucleoid
d. hemichromatin

473. Which of the following histones interact to form the core histone octamer?

a. H1
b. H2A
c. H2B
d. H3
e. H4
f. both *c* and *d*
g. both *d* and *e*
h. all except *a*
i. all of the above

474. What is the net charge of a histone protein?

a. positive
b. negative
c. neutral
d. none of the above

475. What is the overall charge of a nucleosome?

a. positive
b. negative
c. neutral
d. none of the above

476. When DNA is coiled around the histone octamer, which of the following is *true*?

a. It follows a smooth, circular path.
b. The DNA enters and exits on opposite sides of the nucleosome.
c. DNA sequences 40 bp apart are brought into proximity.
d. All of the above are true.
e. None of the above are true.

477. Which of the following is/are true about DNase I hypersensitive sites?

a. They are approximately 100 times more susceptible to digestion by DNase I than normal chromatin.
b. They are regions where the DNA is free of nucleosomes.
c. They usually occur when a gene is being expressed.
d. all of the above

478. Which of the following is/are true when a gene is active?

a. The promoter is free of nucleosomes.
b. The promoter is covered by phased nucleosomes.
c. The histones are typically not acetylated.
d. all of the above

479. Which of the following transcription factors contain TBP?

a. TFIIIA
b. TFIIIB
c. TFIID
d. SL1
e. UBF1
f. both *b* and *c*
g. both *c* and *d*
h. answers *b*, *c*, and *d*
i. all of the above

480. Which of the following transcription factors can be termed *assembly* factors?

a. TFIID
b. TFIIH
c. TFIIIC
d. SP1

481. Phosphorylation of the CTD of RNA Pol II correlates with which of the following?

a. binding to the preinitiation complex
b. the kinase activity of TFIIH
c. the switch from an initiating polymerase to an elongating polymerase
d. both *b* and *c*

482. Which of the following proteins behaves as the eukaryotic counterpart to the bacterial sigma factor?

a. glucocorticoid receptor
b. TFIIIA
c. TBP
d. ERCC1

483. Which RNA polymerase transcribes the 5S ribosomal RNA?

a. RNA Pol I
b. RNA Pol II
c. RNA Pol III
d. telomerase

484. A response element consists of two palindromic half sites, each six bases long. This suggests that the protein that recognizes this element does what?

a. binds as a monomer
b. binds as a dimer
c. contains a zinc finger
d. contains a leucine zipper motif

485. Which of the following is *true* about splicing in *trans*?

a. It never occurs.
b. It can occur if the intron sequences are complementary.
c. It uses an autocatalytic mechanism.
d. It occurs only in yeast tRNA.

486. The λ locus in the human genome contains which of the following?

a. 2 V gene segments
b. ~300 V gene segments
c. 4 J-C gene segments
d. 15 D gene segments

487. Which of the following enzymes does *not* have 5′-3′ exonuclease activity?

a. *E. coli* DNA polymerase I
b. Klenow DNA polymerase
c. *Taq* polymerase
d. none of the above

488. Which of the following activities is a "proofreading" activity?

a. DNA ligase activity
b. 5′-3′ exonuclease activity
c. 3′-5′ exonuclease activity
d. subtilisin activity

489. Reverse transcriptase often has which of the following activities associated with it?

a. RNase H
b. RNase A
c. RNase P
d. RNA polymerase

490. Which of the following enzymes is often used for footprinting of DNA-protein interactions?

a. T4 DNA polymerase
b. MMLV reverse transcriptase
c. DNase I
d. S1 nuclease

491. In order for DNA ligase to act, which of the following is required?

a. 5′-OH
b. 3′-$PO_4$
c. 5′-$PO_4$
d. dNTPs

492. You wish to transfer the $\gamma$-$PO_4$ from ATP to the 5′-end of an oligonucleotide. Which enzyme would you likely use?

a. Klenow DNA polymerase
b. terminal deoxynucleotidyl transferase
c. T4 polynucleotide kinase
d. alkaline phosphatase

493. Which of the following cloning vectors is capable of accepting the largest cloned fragment of DNA?

a. BAC
b. mid
c. phagemid
d. plasmid
e. cosmid

494. Blue-white screening is used to identify colonies containing a recombinant plasmid. In this procedure, disruption of the _________________ gene in the plasmid leads to a _________________ colony.

a. ampicillin resistance, white
b. β-galactosidase, blue
c. X-gal, white
d. β-galactosidase, white

495. In the PCR process, a template is required. The template typically consists of which of the following?

a. dsDNA only
b. ssDNA only
c. dsDNA + annealed primer
d. ssDNA + annealed primer

496. Which of the following is the sequence of steps in a PCR cycle?

a. melting, annealing, extension
b. melting, extension, annealing
c. annealing, melting, extension
d. extension, annealing, melting

497. "A change in DNA sequence that either creates or abolishes a restriction enzyme recognition site"—this definition describes which of the following?

a. SSLP
b. RFLP
c. SNP
d. radiation hybrid

498. In random sequence DNA, what is the probability that a given six-base sequence will match the recognition sequence of a particular restriction enzyme?

a. 1 in 6
b. 1 in 50
c. 1 in 4096
d. 1 in $10^6$

499. SSLPs include which of the following?

a. minisatellites
b. dingle nucleotide polymorphisms
c. restriction fragment length polymorphisms
d. sequence tagged sites

500. In the production of a radiation hybrid, human DNA is damaged by high dose x-rays and the nuclei are then fused to hamster cell nuclei. How is this accomplished?

a. by treatment with HAT media
b. by treatment with Sendai virus
c. by treatment with mouse mammary tumor virus
d. by treatment with DNA ligase

501. Which of the following are methods that can be used to study gene expression?

a. Southern blot
b. Northern blot
c. RT-PCR (reverse transcriptase PCR)
d. RNase protection assays
e. answers *b*, *c*, and *d*
f. all of the above

502. *In situ* hybridization is a technique used by molecular biologists to study gene expression in cells. For this purpose, a single-stranded RNA probe is hybridized to what?

a. DNA found in the nuclei of cells
b. DNA found in the cytoplasm of cells
c. RNA found in the nuclei of cells
d. RNA found in the cytoplasm of cells
e. all of the above

503. What is the major advantage to *in situ* hybridization over other methods of studying gene expression?

a. It enables the identification of a specific gene within an agarose gel.
b. It enables the identification of a specific mRNA transcript within a gel.
c. It enables the identification of cells where a specific protein is found.
d. It enables the identification of the specific cells where a gene is expressed within a tissue.

504. True or false: *In situ* hybridization's major advantage over other methods of studying gene expression is that the histological information is maintained and specific cells that contain the transcript of interest can be identified.

505. Riboprobes are typically used in *in situ* hybridization experiments. What is a riboprobe?

a. a dsDNA probe
b. a ssDNA probe
c. a dsRNA probe
d. a ssRNA probe

506. RNA interference (RNAi) is a post-transcriptional gene silencing method used to study what?

a. the effect of eliminating a gene from the genome of the target cell
b. the effect of preventing transcription of a specific gene in a cell
c. the effect of blocking translation of a specific gene transcript in a cell
d. the effect of interfering with specific gene replication in a cell

507. RNA interference involves adding exogenous nucleic acid to a cell or organism. This nucleic acid is which of the following?

a. dsDNA
b. ssDNA
c. dsRNA
d. ssRNA
e. any of the above

508. In RNA interference, the exogenous nucleic acid that enters the cell is cleaved into short fragments that are 21–25 base pairs in length. What is the enzyme called that performs this cleavage?

a. splicer
b. cutter
c. dicer
d. ribonuclease A

509. In RNA interference, a group of proteins binds to the exogenous nucleic acid once it is cleaved into small pieces and forms a complex referred to as which of the following?

a. DISC (DNA-induced silencing complex)
b. RISC (RNA-induced silencing complex)
c. PISC (protein-induced silencing complex)
d. RIC (RNA-interfering complex)

510. In RNA interference, the silencing complex moves to the target mRNA and does what?

a. cleaves it, thereby preventing its translation
b. completely digests it to individual nucleotides
c. binds it to prevent it from being translated
d. blocks transcription

511. siRNA stands for what?

a. silent interfering RNA
b. strong interfering RNA
c. short or small interfering RNA
d. single interfering RNA

512. Compared to RNAi, siRNA is __________________ effective for work with mammalian cells.

a. less
b. more
c. equally
d. not

513. RNAi or siRNA experiments are used as a faster, less expensive, and more efficient alternative to which of the following?

a. gene knockouts
b. gene recombination
c. RNase protection assays
d. Western blots

514. RNase protection assays are used to study gene expression by providing an estimate of the amount of a particular mRNA that is present in a population of cells (usually a particular tissue or cell type). RNase protection utilizes a nucleic acid probe to hybridize to the target mRNA. What are these probes?

a. dsDNA that matches a short region of the target mRNA
b. ssDNA that has a sequence the same as the mRNA
c. dsRNA that forms an interfering complex
d. ss RNA with the same sequence as the target RNA
e. ssRNA with a sequence complementary to the target RNA

515. RNase protection assays work because the probe hybridizes to the target, and then all of the unhybridized mRNA and unhybridized probe is degraded with which of the following enzymes?

a. EcoRI
b. DNase I
c. ribonuclease P
d. ribonuclease A

516. In RNase protection assays, once the unhybridized RNA is degraded, the remaining RNA is ______________________________.

a. run on a gel and quantified
b. sequenced and analyzed by BLAST
c. run on a Western blot and bound by a labeled antibody
d. quantified in a spectrophotometer

517. In an RNase protection experiment, one generally compares mRNA samples isolated from which of the following?

a. different tissues at the same developmental time point
b. the same tissue at different developmental time points
c. the same tissue treated with different environmental conditions (e.g., a drug)
d. all of the above
e. *a* and *b*

518. DNA fingerprinting analysis is an extremely useful technique now commonly used in molecular forensics. In DNA fingerprinting analysis, PCR is used to amplify DNA sequences found in the human genome. What are these sequences referred to as?

a. RFLPs (restriction fragment length polymorphisms)
b. LINEs (long interspersed repeats)
c. SINES (short interspersed repeats)
d. STRs (short tandem repeats)
e. LTRs (long tandem repeats)

519. In one type of DNA fingerprinting, 13 pairs of primers are used to amplify 13 repeated regions of DNA in the human genome. When more than one sequence is amplified in a PCR reaction using multiple pairs of primers, what is the reaction referred to as?

a. multiplex PCR
b. hot-start PCR
c. poly-PCR
d. combinatorial PCR
e. confusing PCR

520. In DNA fingerprinting, the products of a specific PCR reaction amplified from one pair of primers differ between individuals. Such variation seen between individuals is referred to as which of the following?

a. balanced polymorphism
b. sequence-length polymorphism
c. random polymorphism
d. variable gene transcript (VGT)

521. In DNA fingerprinting, the PCR products must be run on an automated DNA sequencing instrument that can accurately measure the length of the PCR products very precisely. The instrument makes these PCR products visible because the PCR products are what?

a. labeled with a radioactive primer
b. labeled with a biotin-labeled primer
c. labeled with a fluorescent primer
d. labeled by random priming
e. labeled by nick translation

522. Each individual has a unique DNA fingerprint because every individual is different ______________________________.

a. at each of the 13 amplified sites
b. at some of the 13 amplified sites
c. at only one of the amplified sites
d. at 9 of the 13 amplified sites

523. The technique known as 5′ RACE (rapid amplification of cDNA ends) is used in eukaryotes to do what?

a. initiate transcription at the 5′ end of a gene
b. initiate translation at the 5′ end of a transcript
c. identify the 5′ end of a transcript
d. identify the 5′ end of a short or small interfering (si) RNA

524. Nuclear runoffs are used to measure which of the following?

a. the amount of a specific mRNA that was actively being transcribed at a specific time point
b. the amount of total RNA that was actively being transcribed at a specific time point
c. the amount of a specific mRNA that is found accumulated in cells
d. the amount of total RNA that is found accumulated in cells

525. Microarrays (gene chips) are small glass slides that have picoliter quantities of DNA representing (usually) every known gene in the genome of an organism. Each DNA sample representing an individual gene is spotted on the array at a unique location. The spots of DNA representing each gene are what?

a. 20–25 long oligonucleotides
b. 50–75 long oligonucleotides
c. cDNA from each known gene
d. Items *a*, *b*, and *c* are all used in different types of arrays.
e. Items *a* and *b* are used in different types of arrays.

526. In a typical microarray experiment, gene expression in one group of cells is compared to gene expression in a second group of cells. For example, sample A might be normal cells while sample B might be cells of the same type that have become cancer cells. After mRNA is isolated from sample A and sample B, the two mRNA samples are then converted to a single-strand of cDNA using which of the following enzymes?

a. DNA polymerase I
b. Taq polymerase
c. polyA polymerase
d. reverse transcriptase

527. In the microarray experiment described in the question above, the cDNA samples would most commonly be labeled with what?

a. radioactive nucleotides
b. fluorescent dyes
c. biotin
d. digoxigenin
e. any of the above

528. In the microarray experiment described above, four arrays would be used per experiment. One array would be hybridized with the labeled cDNA from sample A, while the second array would be hybridized with the labeled cDNA from sample B. The sample A probe would be labeled with one dye (e.g., Cy3), while the sample B probe would be labeled with a second dye (e.g., Cy5). Typically, a "dye flip" experiment is also done, where the sample A probe is labeled with the second dye (in this case Cy5), while the sample B probe is labeled with the first dye (in this case Cy3). Thus, there would be four labeled probe samples (sample A-Cy3, sample A-Cy5, sample B-Cy5, and sample B-Cy3). Each of the four labeled probe samples is then hybridized to an identical microarray. The results obtained are read by which of the following?

a. spectrophotometer
b. mass spectrometer
c. confocal fluorescence imager
d. spectrofluorimeter

529. In the microarray experiment described above, a spot that hybridizes with probe A but not probe B indicates a gene that is which of the following?

a. expressed in both the cancer and normal cells
b. expressed in the normal cells but not the cancer cells
c. expressed in the cancer cells but not the normal cells
d. not expressed in either cell type

530. In the microarray experiment described above, a spot that hybridizes with probe B but not probe A indicates a gene that is which of the following?

a. expressed in both the cancer and normal cells
b. expressed in the normal cells but not the cancer cells
c. expressed in the cancer cells but not the normal cells
d. not expressed in either cell type

531. In the microarray experiment described above, a spot that hybridizes with both probe A and probe B indicates a gene that is which of the following?

a. expressed in both the cancer and normal cells
b. expressed in the normal cells but not the cancer cells
c. expressed in the cancer cells but not the normal cells
d. not expressed in either cell type

532. In the microarray experiment described above, a spot that hybridizes with neither probe A nor probe B indicates a gene that is which of the following?

a. expressed in both the cancer and normal cells
b. expressed in the normal cells but not the cancer cells
c. expressed in the cancer cells but not the normal cells
d. not expressed in either cell type

533. In the Affymetrix system for producing microarrays, the oligonucleotides are synthesized directly on the glass chip. What is the method of synthesis referred to as?

a. a primer extension
b. a nick translation
c. a random priming
d. a photolithography
e. a polymerase chain reaction

534. Suppose a gene has three exons (exon 1, exon 2, and exon 3) and two introns (intron 1 and intron 2). If this gene is alternatively spliced, how many different transcripts are possible?

a. one
b. two
c. three
d. four
e. eight

535. What is the molecular machine that is responsible for splicing pre-mRNA to the final mRNA?

a. the ribosome
b. the splicesome
c. the centrosome
d. the enzyme splicease

536. In eukaryotes, in order to be translated, a transcript must what?

a. have its introns spliced out and its exons ligated together
b. have a capping nucleotide added to the 5′ end
c. have a poly-A tail added to the 3′ end
d. move from the nucleus to the cytoplasm
e. all of the above

537. The polymerase chain reaction has been applied to the detection of many infectious agents in humans and in the disease-carrying vectors (i.e., mosquitoes, flies, ticks, etc). In order to use PCR to diagnose an infection in a human, a blood sample is often needed (in cases where the infectious agent can be found in the blood). As an example, an effective PCR test is available for the detection of malaria parasites in blood. DNA is purified from a blood sample and a few microliters of the purified sample are used in the PCR test. What component of the PCR provides the species-specificity of the reaction?

a. the DNA polymerase
b. the nucleotide mix
c. the primers
d. the magnesium
e. the buffer

538. PCR is extremely useful in diagnosis of infection because of its great sensitivity. What is the most serious drawback of using PCR for diagnosis of infection?

a. The great sensitivity of the reaction makes false positive results a concern.
b. The lack of specificity of the reaction makes false positive results a concern.
c. The reaction is too slow, making it impractical for clinical use.
d. The reaction is too expensive, making it impractical for clinical use.

539. PCR can be performed on DNA samples isolated from extremely old specimens (ancient DNA) or from specimens that have been preserved in a variety of ways. Why can PCR work on old, degraded DNA samples?

a. because the Taq polymerase is just as active on degraded DNA as on intact DNA
b. because the primers are designed to amplify a very short piece of DNA
c. because two different polymerases are used to give a synergistic amplification
d. because ligase can be used to join the DNA into larger fragments before the PCR

540. Ancient DNA has been successfully isolated from which of the following samples?

a. the "iceman" found frozen in the Austrian Alps
b. woolly mammoths found frozen in the arctic tundra
c. dinosaur blood found in insects preserved in amber
d. items *a*, *b*, and *c*, above
e. items *a* and *b* above

541. For forensic analysis and subsequent PCR, DNA can be successfully isolated from which of the following?

a. dried blood collected at a crime scene
b. semen collected from a rape victim
c. cigarette butts found at a crime scene
d. pieces of bone
e. all of the above
f. items *a*, *b*, and *c*, above

542. True or false: Cancer tissue has been collected for decades in hospitals all over the world as preserved tissue specimens. This tissue represents a tremendous resource for studying the molecular basis of cancer because DNA for use in PCR can be isolated from small amounts of tissue.

543. "Long PCR" can be used to amplify regions of DNA up to 20 or even 40 kb. This is in sharp contrast to regular PCR, where the practical limit is 5 or 6 kb. What is the difference in "long PCR" that is responsible for the tremendous increase in the size of DNA that can be amplified?

a. Long PCR runs for twice as many cycles as normal PCR.
b. Long PCR uses a specially modified thermal cycler.
c. Long PCR uses two different DNA polymerases in the reaction.
d. Long PCR uses primers that are twice as long as the primers used in normal PCR.
e. all of the above

544. The single most important Web site to know for bioinformatics data is the NCBI site that can be found at http://www.ncbi.nlm.gov. The NCBI site has nucleic acid and protein sequence data gathered from around the world and shares data with other databases in Europe and Japan. What does NCBI stands for?

a. National Center for Bioinformation and Informatics
b. National Center for BioInformatics
c. National Center for Biotechnology Information
d. National Center for Biotechnology and Informatics

545. BLAST is an extremely important set of programs designed to compare your query sequence against all of the data in NCBI or against specific databases in NCBI (such as only human sequence data). Such comparisons are among the most important ways to identify an unknown gene sequence. What does BLAST stands for?

a. Basic Local Alignment Search Tool
b. Basic Long Alignment Searching Technology
c. Boolean Logic Alignment Searching Technique
d. Bioinformatics Logical Alignment Searching Technique

546. The database dbEST is a special database within NCBI that contains only EST sequence data. ESTs are expressed sequence tags. An expressed sequence tag is which of the following?

a. a short region of genomic DNA sequence from a single DNA sequencing reaction
b. a short region of polymorphic sequence data from a microsatellite marker
c. a short piece of genomic DNA that includes a promoter
d. a short piece of cDNA sequence from a single DNA sequencing reaction

547. The non-redundant database (nr) includes which of the following?

a. all genomic DNA sequences from all organisms
b. all genomic DNA and cDNA sequences from all organisms (excluding EST data)
c. all genomic DNA, cDNA, and EST data from all organisms
d. all eukaryotic DNA sequence data but not prokaryotic data

548. What is an ORF?

a. an open reading frame
b. an optimized RNA finder
c. ab optimized reading function
d. an oligonucleotide reading finder

549. Genefinder programs look for sequences in what?

a. cDNA that look like genes
b. genomic DNA that look like genes
c. cDNA that have a promoter
d. genomic DNA that can bind ribosomes

550. A gene fusion is most often the fusion of which of the following?

a. three different genes into a single plasmid vector
b. two different genes into a single plasmid vector
c. two different genes into a bacteriophage vector
d. a promoter of one gene to the coding sequence of another gene

551. Gene fusions are often used to follow the expression of a gene by fusing the gene's promoter to a "reporter" gene. A reporter gene is a gene that encodes a protein product that is easy to visualize in cells where the gene is being expressed. Thus, expression regulated from the promoter can be followed much more easily than by tracking expression of the gene that the promoter is normally controlling. Examples of commonly used reporter genes include which of the following?

a. green fluorescent protein
b. firefly luciferase
c. beta-galactosidase
d. items *a* and *c*, above
e. items *a*, *b*, and *c*, above

552. An expression vector is typically a plasmid vector that contains regulatory elements necessary to achieve high levels of expression of your gene (i.e., to get protein produced) in your host cells of interest. *E. coli* is most commonly used as the host for such expression, although yeast cells, insect cells, and mammalian cells grown in culture can also be used. To achieve high levels of expression of a human gene in *E. coli*, the plasmid expression vector must contain which of the following?

a. a strong *E. coli* promoter
b. an *E. coli* ribosome binding site
c. restriction sites for cloning the gene of interest directly downstream from the promoter and ribosome binding site
d. items *a* and *b* above
e. items *a*, *b*, and *c*, above

553. Some expression vectors include additional sequences that follow the regulatory sequences on the plasmid but come before the restriction sites where the gene of interest is cloned. These additional sequences encode amino acids that simplify the purification of the recombinant fusion protein that is produced. These proteins are called fusion proteins because they start with amino acids encoded on the vector followed by the amino acids encoded by the cloned gene of interest. One of the most well-known types of expression vector includes what is referred to as a "six His tag." This sequence on the vector encodes six histidine amino acids fused to the cloned sequence of the gene of interest. Therefore, the resulting fusion protein has six histidine amino acids at the N-terminal end of the protein. *E. coli* cells containing this gene fusion are grown and the fusion protein is produced. The cells are then lysed and the lysate is poured over a column that binds only the proteins with the "six His tag." Only the "six His tag" fusion protein binds to the column, while all of the rest of the *E. coli* proteins wash through the column. Once all of the contaminating protein is washed out of the column, the fusion protein of interest is easily eluted. The "six His tag" fusion protein binds to the column because the column contains a resin with which of the following?

a. DEAE cellulose
b. nickel, or possibly other divalent cations such as $Zn^{+2}$ or $Co^{+2}$
c. oligo-dT cellulose
d. sepharose

554. Obtaining high levels of protein expression in *E. coli* using an expression vector can be useful for producing large amounts of the recombinant fusion protein for what purpose?

a. use as a drug (e.g., insulin)
b. production of large quantities of the protein for structural studies (e.g., x-ray crystallography)
c. use as a subunit vaccine
d. answers *a* and *c*, above
e. answers *a*, *b*, and *c*, above

555. For genomics research, large fragment DNA libraries are often constructed in vectors that allow the cloning of DNA fragments that are 100 kb in length or greater. These large fragments simplify the assembly of the DNA sequence of the entire genome. One commonly used type of vector for constructing such large fragment libraries are "Bacterial Artificial Chromosomes" or BACs. These BAC vectors are cleaved with a restriction endonuclease, and large fragments of the genome of interest are ligated into the vector. Fragments are typically 80 kb to 150 kb in length. The recombinant vectors are then transformed into *E. coli* cells where it can replicate. The BAC vector is really a plasmid vector with a strict copy number control mechanism such that only one copy of the BAC is found per *E. coli* cell. The advantage of this strict copy number control is that it limits recombination of the large insert DNA. Such recombination can create major problems in the assembly of the sequence of the genome. In order to initiate construction of a BAC library, the genome of interest must be sheared or cleaved into very large fragments in the 100 kb range. Fragments this large cannot be separated on traditional agarose gels. In order to measure the size of the cleaved genomic DNA and to purify DNA in the right size range, one can use which of the following?

a. pulsed field gel electrophoresis
b. polyacrylamide gel electrophoresis
c. CsCl gradient ultracentrifugation
d. sucrose gradients

556. Once the genomic DNA has been cleaved and the proper fragment sizes isolated, the fragments are ligated into the BAC vector. The very large recombinant DNA molecules must then be transferred into *E. coli* cells using a very efficient technique (because it is so difficult to get large molecules into *E. coli* cells). The most common technique used with BAC vectors is which of the following?

a. pulsed field electrophoresis
b. transformation of competent cells using a heat shock
c. electroporation
d. infection with bacteriophage lambda
e. lipofection

557. Because the recombinant BACs are so difficult to get into the *E. coli* cells, there must be an easy way to select for those *E. coli* cells that have taken up a BAC vector from the vast majority of cells that have not. This is easily accomplished because the BAC vectors contain at least one antibiotic resistance gene. This selection is referred to as what?

a. the selection for recombinants
b. the selection for transformants
c. blue/clear selection
d. the selection for expression

558. The collection of BAC clones from the experiment described above will yield a large fragment library containing very large pieces of your genome of interest. This BAC library can now serve as an important resource for the assembly of a complete genomic sequence. In addition to this large fragment library, it is also useful to have smaller fragment libraries constructed in plasmids. In recent years, it has been found to be very efficient if a library is constructed with an average insert size of 10 kb and another library with an average insert size of 2 kb. These small fragment libraries serve as the templates for most of the genomic sequencing in a genome sequencing project. In order to obtain DNA fragments of the appropriate size for each library, researchers do what?

a. cleave the DNA with an 8-cutter restriction enzyme (an enzyme that recognizes an 8-base-pair-long restriction site)
b. cleave the DNA with the enzyme DNaseI
c. randomly shear the DNA by nebulization
d. cleave the DNA with a 4-cutter restriction enzyme

559. DNA sequence data obtained from sequencing the ends of the BAC clones and the smaller insert plasmid clones are all used in assembling the complete genomic sequence of an organism. In order to obtain the complete sequence of the genome, a large amount of DNA must be sequenced. The size of the human genome is 3 billion base pairs. Approximately how many base pairs of genomic sequence need to be sequenced to assemble the entire sequence of the human genome?

a. 3 billion base pairs
b. 6 billion base pairs
c. 12 billion base pairs
d. 30 billion base pairs

560. Once the "complete" genome sequence is obtained, the work of "finishing" the genome sequence must be done. This is because gaps often remain in the sequence that prevent the final assembly of the genome. When President Clinton, J. Craig Venter (then the president of Celera, the private company that sequenced the human genome), and Francis Collins (the director of the public genome project) joined in making their famous announcement in 2000, was the human genome sequence work finished?

a. yes
b. no

561. Once the sequencing of a genome is complete, there is still much work that needs to be done to make the sequence data as useful as possible for the scientific community. First, the genes must be located on the DNA using all available evidence. This is referred to as gene modeling. Gene modeling uses all data that can be helpful in identifying genes in the genomic sequence. The data used includes which of the following?

a. cDNA sequences and EST data
b. genefinder programs
c. gene sequences identified from closely related species
d. all of the above

562. Once the genes have been located on the genome sequence, the genes need to be annotated in the database. Gene annotation includes which of the following?

a. a tentative identification of the function of the product encoded by the gene based on BLAST and other comparative analyses
b. a complete description of the regulatory sequences (promoter, enhancer, etc.) that control expression of the gene
c. a complete description of the three-dimensional structure of the encoded protein
d. all of the above

563. YAC vectors (Yeast Artificial Chromosomes) enable the cloning of very large fragments of DNA into a vector that can replicate in both yeast cells and in *E. coli*. These YAC vectors have which of the following sequences?

a. a cloned yeast centromere
b. a cloned yeast ARS (autonomously replicating sequence)
c. two telomere sequences
d. yeast selectable markers (e.g., URA3)
e. an *E. coli* plasmid origin of replication
f. an antibiotic resistance gene
g. all of the above
h. only *a*, *c*, *d*, and *f*, above

564. Cosmids are vectors that can accept inserts up to approximately 40 kb in length. Why can the recombinant cosmids be introduced into *E. coli* cells with great efficiency?

a. because they can be packaged in lambda bacteriophage heads and tails and the resulting viable phage can be used to infect *E. coli*
b. because they are introduced into *E. coli* using electroporation, which is more efficient than using bacteriophage lambda
c. because they are introduced into *E. coli* by lipofection, a technique that is very efficient for introducing DNA into *E. coli*
d. because they are introduced into *E. coli* using calcium phosphate precipitation with a heat shock
e. All of the above are equally efficient for introducing DNA into *E. coli*.

565. Which of the following is the reason that *E. coli* cells don't "commit suicide" by cutting their own DNA with the *Eco*RI restriction enzyme?

a. *E. coli* does not produce *Eco*RI.
b. *E. coli* does not have any *Eco*RI restriction sites in its genome.
c. *E. coli* produces an *Eco*RI methylase enzyme that also recognizes an *Eco*RI site. The enzyme methylates the DNA at those sites and protects them from cutting by *Eco*RI.
d. *E. coli* produces *dam* methylase which protects the DNA from digestion by *Eco*RI.
e. none of the above

566. Why do bacterial cells possess restriction/methylation genes?

________________________________________

567. In eukaryotic cells, histone proteins are known to play a crucial role in chromosome structure (nucleosomes). However, histone proteins are also thought to play an important role in which of the following?

a. regulation of DNA replication
b. regulation of transcription
c. regulation of translation
d. regulation of RNA processing
e. all of the above

568. What is an operon?

a. a cluster of genes found together by random chance in the genomes of bacteria
b. a cluster of functionally related genes under the control of a single promoter
c. a cluster of functionally related genes that are transcribed together and then translated into a large polyprotein that is cleaved into separate proteins following translation
d. a cluster of functionally related genes found together on the genome but regulated independently

569. The *lac* operon of *E. coli* is an example of the elegant yet simple methods developed by bacteria to regulate their genes. What are the genes of the *lac* operon involved in?

a. the synthesis of the sugar lactose
b. the breakdown of the sugar lactose
c. the transport of the sugar lactose into the cell
d. the export of the sugar lactose out of the cell
e. both *b* and *c*
f. both *a* and *d*

570. The genes of the *lac* operon include which of the following?

a. the beta-galactosidase gene (*lac*Z)
b. the lactose permease gene (*lac*Y)
c. the acetylase gene (*lac*A)
d. the *lac* repressor gene (*lac*I)
e. items *a*, *b*, and *c*, above
f. all of the above.

571. The *lac*I gene encodes the lac repressor protein, which binds to the ____________________ to shut down transcription of the lac operon.

a. *lac* operator
b. *lac* promoter
c. *lac* terminator
d. *lac* repressor gene

572. The *lac* repressor is a regulatory protein that binds to the *lac* operon to shut off transcription ________________________________.

a. when lactose is absent.
b. when lactose is present.
c. whether or not lactose is present.
d. when lactose is unable to enter the cell.

573. The regulation of the *lac* operon by the presence or absence of lactose is referred to as what?

a. positive regulation
b. negative regulation
c. inducible regulation
d. repressible regulation
e. both *a* and *d*
f. both *b* and *c*
g. both *a* and *c*
h. both *b* and *d*

574. In addition to the regulation by lactose and the *lac* repressor protein, there is another completely separate form of regulation in the *lac* operon. This regulation involves glucose and is often referred to as the "glucose effect." The basis of this regulation is that when both glucose and lactose are present, glucose is the preferred carbon and energy source for *E. coli*. When both sugars are present, the *E. coli* cells will first use the glucose and then the lactose. Which of the following statements about this form of regulation is *true*?

a. When glucose levels in the cell are high, glucose binds to the *lac* repressor that binds to the *lac* operator and blocks transcription of the operon.
b. When glucose levels in the cell are high, glucose releases from the *lac* repressor, which causes it to lose its affinity for the operator, resulting in transcription of the operon.
c. When glucose levels in the cell are low, cAMP levels rise, and cAMP binds to the CRP protein that then binds to the *lac* promoter, increasing transcription of the operon.
d. When glucose levels in the cell are low, cAMP levels are also low; this enables CRP to be free of cAMP and bind to the promoter to increase transcription of the operon.

575. Why does glucose inhibit production of cAMP?

a. It inhibits the enzyme adenyl cyclase, which converts ATP to cAMP.
b. It inhibits ATPase, which prevents the production of cAMP.
c. It inhibits the conversion of AMP to cAMP by the enzyme AMPase.
d. It blocks transcription of the gene-encoding CRP protein.

576. Why does it make sense in *E. coli* for glucose to inhibit transcription of the *lac* operon?

a. Because *E. coli* can't use both glucose and lactose at the same time. This would be metabolically impossible.
b. Because *E. coli* can use glucose more efficiently than lactose. Therefore, it makes sense to use the glucose first and then the lactose.
c. Because in *E. coli*, if glucose is present, transcription of all genes is repressed.
d. Because in *E. coli*, if there is no glucose present, all operons are expressed at high levels.

577. The regulation of the *lac* operon by the presence or absence of glucose is referred to as what?

a. positive regulation
b. negative regulation
c. neutral regulation
d. attenuation

578. The *lac* operon is an operon in which the genes are involved in the metabolism of the sugar lactose. Other sugar operons, such as arabinose and maltose, are also inducible; i.e., presence of the sugar induces expression of the operon. On the other hand, operons involved in the biosynthesis of amino acids such as tryptophan are referred to as which of the following?

a. negative
b. positive
c. inducible
d. repressible

579. In the *trp* operon of *E. coli* the small effector molecule tryptophan is referred to as which of the following?

a. the repressor
b. the co-repressor
c. the inducer
d. the co-inducer

580. In the *trp* operon, the trp repressor binds to which DNA sequence to repress transcription of the *trp* operon?

a. the promoter
b. the attenuator
c. the operator
d. the terminator

581. Another form of regulation in the *trp* operon is regulation that involves translation of a short region of the *trp* operon that precedes the first gene in the operon. This type of regulation is referred to as what?

a. promotion
b. induction
c. repression
d. attenuation

582. Transcription of the *trp* operon is attenuated when ____________________.

a. tryptophan levels in the cell are low
b. tryptophan levels in the cell are high
c. glucose levels in the cell are high
d. glucose levels in the cell are low
e. both *b* and *c*
f. both *a* and *d*

583. The original version of the Ames test to screen for mutagens and cancer suspect agents utilizes which of the following?

a. special strains of *Escherichia coli*
b. special strains of *Salmonella typhimurium*
c. special strains of *Helicobacter pylori*
d. special strains of *Bacillus subtilis*

584. The original version of the Ames test relies on the reversion of which of the following?

a. his- missense mutations to his+
b. arg- missense mutations to arg+
c. his- frameshift mutations to his+
d. arg- frameshift mutations to arg+
e. both *a* and *c*
f. both *b* and *d*
g. both *a* and *b*
h. both *c* and *d*

585. In a hypothetical Ames test experiment to check for the mutagenicity of condensates from cigarette smoke, the following results were obtained:

| | |
|---|---|
| Bacterial missense strain + pure water | 12 revertant colonies |
| Bacterial missense strain + smoke condensate | 9 revertant colonies |
| Bacterial frameshift strain + pure water | 0 revertant colonies |
| Bacterial frameshift strain + smoke condensate | 117 revertant colonies |

Based on the above results, which of the following conclusions seem(s) reasonable?

a. Water is a missense mutagen.
b. Water is a frameshift mutagen.
c. Cigarette smoke is a missense mutagen.
d. Cigarette smoke is a frameshift mutagen.
e. Water is both a missense and a frameshift mutagen.
f. Cigarette smoke is both a missense and a frameshift mutagen.

586. Three routes for the introduction of DNA into bacterial cells include which of the following?

a. transcription, translation, and transformation
b. transcription, transformation, and conjugation
c. transformation, conjugation, and translation
d. transformation, conjugation, and transduction
e. transformation, transduction, and translation

587. Transformation is the process by which naked DNA is taken up by bacterial cells either naturally (as in *Bacillus* and *Hemophilus*) or with a laboratory treatment as with *E. coli*. In order to transform *E. coli* cells, they must be made ______________________________.

a. conjugative
b. resistant
c. transposable
d. transducible
e. competent

588. Bacterial conjugation is a process that enables DNA to pass from one cell to another. Bacterial conjugation is also often referred to as what?

a. transduction
b. transposition
c. bacterial mating
d. bacteriophage infection

589. Bacterial conjugation in *E. coli* requires genes that are encoded on which of the following?

a. R plasmid
b. F plasmid
c. E plasmid
d. C plasmid

590. Genes on the fertility factor encode which of the following?

a. the proteins that make up the sex pilus
b. the proteins that make up the resistance pilus
c. the proteins that make up the transposon
d. the proteins that make up the transducing phage

591. Plasmids often have genes that encode which of the following?

a. toxins
b. bacteriocins
c. antibiotic resistance
d. enzymes involved in the metabolism of unusual sugars and hydrocarbons
e. all of the above
f. items *a* and *c*, above
g. items *a*, *b*, and *c*, above

592. What are R factors?

a. plasmids that carry genes that encode resistance to infection by a bacteriophage.
b. plasmids that carry antibiotic resistance genes.
c. plasmids that carry recombination genes.
d. plasmids that carry DNA repair genes.
e. plasmids that carry regulatory genes.

593. R factors (R plasmids) were first discovered________________________.

a. in Southeast Asia in the late 1960s from *Salmonella* species
b. in Europe in the early 1970s from toxic *Escherichia* species
c. in Japan in the late 1950s after dysentery epidemics
d. in Australia in the mid-1960s from *Klebsiella* species

594. What is horizontal gene transfer?

a. the transfer of genes from one species to another
b. the transfer of genes from one cell to another cell of the same bacterial species *via* plasmids
c. the transfer of genes from one cell to another cell of the same species via transduction
d. all of the above

595. Transposons are which of the following?

a. a key element of transcription
b. a key element of translation
c. a key element of transduction
d. short segments of DNA that can move from one position to another

596. Transposons occur in what?

a. in prokaryotes only
b. in eukaryotes only
c. in both prokaryotes and eukaryotes
d. only in archaea

597. In what were transposons were first discovered in the 1950s by Barbara McClintock?

a. *Drosophila*
b. *E. coli*
c. corn
d. mice
e. humans

598. In prokaryotes, the simplest transposons possess only a gene that encodes a transposase enzyme and inverted repeat sequences. What are these simple transposons referred to as?

a. transferases
b. transductons
c. transfer elements
d. insertion sequences

599. Complex transposons include other genes not involved in transposition. These genes include which of the following?

a. antibiotic resistance genes
b. enterotoxin genes
c. neither of the above
d. both *a* and *b*

600. Transposons can move from a location on the chromosome to where?

a. another position on the chromosome
b. a plasmid
c. to a bacteriophage DNA
d. all of the above

601. Transduction is the process by which ______________________________.

a. genetic material is transferred from one bacterial cell to another *via* a bacteriophage
b. genetic material is transferred from one bacterial cell to another *via* conjugation
c. genetic material is transferred from one bacterial cell to another *via* transformation
d. all of the above

602. DNA was proven to be the genetic material in 1944 by Avery, MacLeod, and McCarty when they worked with *Streptococcus pneumoniae*. In these experiments they were able to do what?

a. transform healthy mice to mice that died from pneumonia
b. transform avirulent bacteria to virulent bacteria
c. transform nonencapsulated bacteria to bacteria with a protective capsule
d. all of the above
e. only *b* and *c*

603. In bacterial mating (conjugation) an Hfr cell is a cell that has what?

a. the F factor (F plasmid) carrying a piece of bacterial chromosomal DNA
b. the F factor (F plasmid) carrying a piece of another plasmid
c. the F factor (F plasmid) carrying a piece of bacteriophage DNA
d. the F factor (F plasmid) integrated into the bacterial chromosome
e. all of the above
f. only *b* and *c*

604. In bacterial mating (conjugation), an F- cell is a cell that has what?

a. the F factor (F plasmid) carrying a piece of bacterial chromosomal DNA
b. the F factor (F plasmid) carrying a piece of another plasmid
c. the F factor (F plasmid) carrying a piece of bacteriophage DNA
d. the F factor (F plasmid) integrated into the bacterial chromosome
e. no F factor

605. In bacterial mating (conjugation) an F′ cell is a cell that has what?

a. the F factor (F plasmid) carrying a piece of bacterial chromosomal DNA
b. the F factor (F plasmid) carrying a piece of the bacterial chromosome
c. the F factor (F plasmid) carrying a piece of bacteriophage DNA
d. the F factor (F plasmid) integrated into the bacterial chromosome
e. all of the above

606. What happens to the F- cell when an F+ male *E. coli* cell mates with an F- female *E. coli* cell?

a. It is converted to an F+ cell.
b. It is converted to an F′ cell.
c. It is converted to an Hfr cell.
d. It remains an F- cell.
e. none of the above

607. Why is passage of plasmids from one cell to another *via* bacterial conjugation an important public health problem?

a. because it produces Hfr recombinant cells that produce novel antibiotics
b. because it produces recombinant antibiotics
c. because it results in the rapid spread of antibiotic resistance genes
d. none of the above
e. all of the above

608. Genetically engineered cells can be used to produce drugs useful for treating human disease. Which of the following types of cells have been used successfully for producing such recombinant drugs?

a. *Escherichia coli*
b. *Saccharomyces cerevisiae* (yeast)
c. mammalian cells grown in culture
d. *Pichia pastoris* (yeast)
e. all of the above
f. none of the above

609. Which of the following are examples of pharmaceutical products that are currently produced in *E. coli*?

a. alpha-interferon
b. epidermal growth factor (EGF)
c. human growth hormone (hGH)
d. human insulin
e. tumor necrosis factor (TNF)
f. items *c* and *d* above
g. items *a*, *b*, and *e*, above
h. items *a*, *b*, *c*, *d*, and *e*, above

610. Which of the following are examples of pharmaceutical products that are currently produced in *Saccharomyces cerevisiae*?

a. alpha-interferon
b. hepatitis B vaccine
c. influenza vaccine
d. prourokinase
e. superoxide dismutase (SOD)
f. items *d* and *e*, above
g. items *b*, *c*, and *d*, above
h. items *a*, *b*, *c*, *d*, and *e*, above

611. Which of the following are examples of pharmaceutical products that are currently produced in *Pichia pastoris* (yeast)?

a. alpha-interferon
b. antitrypsin
c. factor VIII
d. relaxin
e. superoxide dismutase
f. items *b*, *c*, and *d*, above
g. items *a* and *e*, above
h. items *a*, *b*, *c*, *d*, and *e*, above

612. Which of the following are examples of pharmaceutical products that are currently produced in mammalian cells in culture?

a. beta-interferon
b. erythropoietin (EPO)
c. factor VIII
d. tissue plasminogen activator
e. pulmozyme (rhDNase)
f. items *b*, *c*, and *d*, above
g. items *a* and *e*, above
h. items *a*, *b*, *c*, *d*, and *e*, above

613. The pharmaceutical product antitrypsin is produced by which of the following?

a. *Escherichia coli*
b. *Saccharomyces cerevisiae*
c. genetically modified sheep
d. mammalian cells in culture
e. *Pichia pastoris*

614. Which of the following vectors have been utilized in gene therapy?

a. adenoviruses
b. retroviruses
c. plasmids
d. bacteriophage lambda
e. items *a* and *b*, above
f. items *a*, *b*, and *c*, above
g. items *a*, *b*, *c*, and *d*, above

615. What does the human *p53* gene encode?

a. a protooncogene
b. a tumor suppressor protein
c. a gene with 53 splice variants
d. a tumor-specific transplantation antigen.

616. What is the estimated number of genes in the human genome?

a. 10,000–15,000
b. 30,000–40,000
c. 75,000–100,000
d. 120,000–150,000

617. What technique was crucial to the identification of *Tropheryma whippelii* as the causative agent of Whipple's disease (a gastrointestinal and nervous system disorder)?

a. Enzyme-linked immunosorbent assay (ELISA)
b. Western blotting
c. polymerase chain reaction (PCR)
d. slide agglutination test

618. What virus was identified by PCR in less than two weeks after an outbreak of hemorrhagic fever in the United States in 1993?

a. Rocky Mountain spotted fever
b. Equine encephalitis virus
c. Legionellosis
d. Hanta virus
e. Cat-scratch disease

619. What bacterium was identified by PCR as the causative agent of a new tick-borne human granulocytic disease in 1994?

a. *Ehrlichia chaffeensis*
b. *Yersinia gastroenteritis*
c. *Borrelia burgdorferi*
d. *Leptospira interrogans*

620. The first PCR test kit for a pathogenic organism received approval from the Food and Drug Administration in 1993. This kit was for the detection of which of the following?

a. *Chlamydia trachomatis*
b. *Neisseria gonorrhoeae*
c. *Listeria monocytogenes*
d. *Streptococcus pneumoniae*

**621–642.** Viruses that infect humans can have genomes that are double-stranded DNA, single-stranded DNA, double-stranded RNA, or single-stranded RNA. The type of genome that a virus possesses is an important criterion used to categorize viruses into families and genera. **For each virus, list its genome type (dsDNA, ssDNA, dsRNA, ssRNA)**.

621. Hepatitis D (Deltaviridae) __________

622. Coronavirus (associated with the common cold) __________

623. Reovirus (one type causes Colorado tick fever) __________

624. Ebola virus (a Filovirus) __________

625. Hepatitis B __________

626. Human parvovirus B19 __________

627. Western equine encephalitis __________

628. Human papillomavirus (HPV) __________

629. Influenza virus __________

630. Hanta virus __________

631. Dengue (causes Dengue fever) __________

632. Smallpox virus __________

633. Hepatitis E __________

634. Lassa fever virus __________

635. Rabies virus __________

636. Cytomegalovirus __________

637. Mumps virus __________

638. West Nile virus __________

639. HIV (a lentivirus) __________

640. Hepatitis A __________

641. Polio virus __________

642. Rhinovirus (another common cold type of virus) __________

643. In 1999, the reverse-transcriptase polymerase chain reaction (RT-PCR) was used to amplify the RNA genome of a virus found in both birds and humans to identify an outbreak of which of the following?

a. Hanta virus
b. equine encephalitis virus
c. West Nile virus
d. Marburg virus
e. all of the above

644. Oncogenic viruses with DNA genomes include which of the following?

a. human papillomaviruses (HPV)
b. Epstein-Barr virus
c. hepatitis B virus
d. HTLV-1.
e. items *a*, *b*, and *c*, above
f. items *a*, *b*, *c*, and *d*, above

645. Oncogenic viruses with RNA genomes include which of the following?

a. human T-cell leukemia viruses
b. feline leukemia virus (FeLV)
c. mouse mammary tumor virus
d. all of the above
e. none of the above

646. What is the causative agent of Creutzfeldt-Jakob disease (CJD)?

a. dsDNA virus
b. ssDNA virus
c. dsRNA virus
d. ssRNA virus
e. none of the above

647. Which of the following human diseases is thought to be caused by prions?

a. kuru
b. scrapie
c. Gerstmann-Straussler-Scheinker syndrome
d. fatal familial insomnia

648. Viroids are small infectious agents that ________________________.

a. are small ssRNA viruses that cause neurological diseases in humans
b. are short pieces of naked RNA with no protein coat that cause some plant diseases
c. are small prion-like particles that cause neurlogical diseases in mammals
d. are very small plant viruses that consist of ssDNA and a small protein coat
e. none of the above

649. In 1985, a giant bacterial species (*Epulopiscium fishelsoni*) was discovered that is about one million times the volume of *E. coli*. These bacteria live symbiotically in the gut of the Red Sea surgeonfish and were originally thought to be protozoans. Why was it later determined that these newly discovered microbes were bacteria?

a. because they had no membrane-enclosed nucleus
b. because they were smaller than the smallest known eukaryotic cells
c. because they have flagella that resemble bacterial flagella
d. because they reproduce by binary fission
e. all of the above
f. items *a* and *c*, above
g. items *a*, *c*, and *d*, above

650. All of life has recently been divided into three "domains" called Eukarya (the eukaryotes), Bacteria, and Archaea (formerly called the archaebacteria). Which of the following statements is/are *true* about the domains Archaea and Bacteria?

a. Bacterial cell walls contain peptidoglycan, while Archaea cell walls do not.
b. Bacteria use formylmethionine as the first amino acid in protein synthesis while Archaea use methionine.
c. Bacteria and Archaea are sensitive to the same antibiotics.
d. All of the above are true.
e. Only *a* and *b* are true.
f. Only *b* and *c* are true.
g. Only *a* and *c* are true.

651. Regarding the domains Archaea and Eukarya, which of the following statements is/are *true*?

a. The membrane lipids of both Archaea and Eukarya are composed of branched carbon chains attached to glycerol by ether linkages.
b. The first amino acid in protein synthesis in both Archaea and Eukarya is methionine.
c. Both Archaea and Eukarya have a common arm of tRNA that has the following sequence: guanine-thymine-pseudouridine-cytosine-guanine.
d. All of the above are true.
e. Only *a* and *b* are true.
f. Only *b* and *c* are true.
g. Only *a* and *c* are true.

652. Pathogenicity islands in bacteria ______________________________.

a. are large regions of DNA that carry genes responsible for virulence
b. have been acquired during evolution by horizontal gene transfer
c. occur in only one position in the genome
d. all of the above
e. items *a* and *b*, above
f. items *a* and *c*, above
g. items *b* and *c*, above

653. One example of a set of genes carried in a pathogenicity island are genes that encode various pathways for protein secretion. An example is the type III secretion system that facilitates the entry of virulence proteins into the cytoplasm of the eukaryotic host cells. These virulence proteins ____________________.

a. directly cause lysis of the host eukaryotic cells
b. interfere with host cell signal transduction
c. bind and inactivate host antibodies
d. all of the above
e. items *a* and *b*, above
f. items *a* and *c*, above
g. items *b* and *c*, above

654. Pathogenicity islands are found in pathogenic members of a genus or species but not in non-pathogenic members of the same genus or species. Which of the following are examples of pathogenic bacteria that carry pathogenicity islands?

a. *Yersinia* spp
b. *Pseudomonas aeruginosa*
c. *Shigella flexneri*
d. *Salmonella typhimurium*
e. enteropathogenic *E. coli*
f. all of the above

655. Which of the following statements is *true* about bacterial toxins?

a. Exotoxins are frequently encoded by genes on plasmids or prophages, while endotoxins are encoded by chromosomal genes.
b. Endotoxins are frequently encoded by genes on plasmids or prophages, while exotoxins are encoded by chromosomal genes.
c. Both exotoxins and endotoxins are frequently encoded by genes on plasmids or prophages.
d. Both exotoxins and endotoxins are always encoded by chromosomal genes.

656. Which of the following statements is *true* about bacterial toxins?

a. Exotoxins are usually secreted outside the bacterial cell, while endotoxins are found in the cytoplasm of the bacteria.
b. Exotoxins are usually secreted outside the bacterial cell, while endotoxins are found as part of the outer membrane of gram-negative bacteria.
c. Exotoxins are secreted onto the surface of the eukaryotic host cell, while endotoxins are secreted into the interior of the eukaryotic host cell.
d. Endotoxins are generally more toxic than exotoxins.

**657–670. Match the antibiotics with the antibiotic mode of action (each mode of action may be used more than once).**

657. Tetracycline _____
658. Polymyxin B _____
659. Penicillin _____
660. Fusidic acid _____
661. Rifampin _____
662. Ampicillin _____
663. Sulfonamides _____
664. Streptomycin _____
665. Ciprofloxacin _____
666. Bacitracin _____
667. Gentamicin _____
668. Erythromycin _____
669. Trimethoprim _____
670. Chloramphenicol ____

Antibiotic modes of action:

a. binds the 30S ribosomal subunit: inhibits protein synthesis, causing misreading of the mRNA
b. binds the 50S ribosomal subunit to block peptide bond formation
c. binds the 30S ribosomal subunit and interferes with aminoacyl-tRNA binding
d. binds the 50S ribosomal subunit and inhibits elongation of the protein chain
e. binds to EF-G and blocks translocation of the ribosome
f. inhibits bacterial DNA gyrase and interferes with DNA replication
g. blocks RNA synthesis by binding/inhibiting RNA polymerase
h. inhibits transpeptidation enzymes and activates cell wall lytic enzymes
i. binds to the plasma membrane and disrupts its structure
j. disrupts some aspect of normal bacterial metabolism
k. inhibits cell wall synthesis by interfering with lipid carriers

671. True or false: R plasmids can carry a variety of different types of antibiotic resistance genes.

One mode of action of antibiotic resistance is via genes that encode enzymes that destroy or modify the antibiotic (e.g., the hydrolysis of penicillin).

**672–679. Match the antiviral drugs with the antiviral mode of action (each mode of action may be used more than once).**

672. AZT (azidothymidine) or zidovudine _____
673. Amantadine ____
674. Acyclovir _____
675. Vidarabine or adenine arabinoside _____
676. Saquinvir, indinavir, and ritonavir _____
677. Interferons _____
678. Foscarnet _____
679. HPMPC or cidofovir _____

Antiviral modes of action:

a. blocks the penetration and uncoating of viral particles
b. disrupts the activity of the viral DNA polymerase
c. blocks cleavage of pyrophosphate from nucleoside triphosphate substrates
d. inhibits the activity of reverse transcriptase
e. inhibits the activity of HIV protease
f. helps mobilize the immune system against the virus

680. Ribotyping is a method of bacterial strain typing that involves which of the following?

a. the isolation of ribosomes and their characterization to distinguish one strain from another
b. the sequencing of ribosomal RNA to distinguish one strain from another
c. cleaving the chromosomal DNA of the bacterial strain with a restriction enzyme and hybridizing that DNA on a Southern blot with rRNA probes
d. classifying the various rRNA molecules found in cells into their various types (i.e., 23S, 16S, and 5S)

681. Plasmid fingerprinting is another method of bacterial strain typing. In this procedure, which of the following is true?

a. Microbial isolates are identified by isolating plasmids and sequencing them.
b. Microbial isolates are identified by isolating plasmids and examining them on agarose gels.
c. Microbial isolates are identified by isolating plasmids and running them on sucrose gradients to determine their molecular weights.
d. Microbial isolates are identified by isolating plasmids and submitting them to the ribotyping protocol.

682. An important exception to the genetic code specifies the so-called 21st amino acid (selenocysteine). This amino acid is specified by the codon UGA, which is usually a stop codon. This exception is found in bacteria, archaea, and eukaryotes. In cases where selenocysteine is found in proteins, it is because of which of the following?

a. The amino acid cysteine is incorporated into the protein and then is modified to selenocysteine post-translationally.
b. The amino acid selenocysteine is incorporated directly into the growing polypeptide chain, but this does not involve a tRNA. Instead, a protein is involved in "bringing" the selenocysteine to the ribosome and inserting it opposite the UGA codon.
c. The amino acid selenocysteine is incorporated directly into the growing polypeptide chain using a tRNA that recognizes the UGA codon under special circumstances.
d. none of the above

683. True or false: The UGA codon also can encode tryptophan in human mitochondria, some yeast, and in mycoplasmas and spiroplasmas.

684. True or false: A 22nd amino acid called pyrrolysine is bound by a tRNA that can recognize the stop codon UAG in some archaea.

685. True or false: The codon AUA (which normally encodes isoleucine) encodes methionine in archaea.

686. True or false: The codon UAG (which is normally a stop codon) codes for glutamine in some protozoa and algae.

687. True or false: The codon CGG (which normally codes for arginine) codes for tryptophan in all mitochondria.

688. True or false: AUG encodes methionine in eukaryotes, but N-formylmethionine in bacteria and archaea.

689. Many *Staphylococcus* species are human pathogens that cause food poisoning, ear infections, and toxic shock syndrome. Many strains are resistant to the antibiotic penicillin because they have a gene that encodes the protein beta-lactamase. How does beta-lactamase render the bacterium resistant to penicillin?

a. It blocks entry of the penicillin into the cell.
b. It causes secretion of the penicillin following its entry into the cell.
c. It cleaves the penicillin
d. It adds a lactam group to the penicillin, thereby inactivating the antibiotic.
e. none of the above

690. Medical researchers hope to make pencillin-resistant *Staphylococcus* sensitive to penicillin by understanding how penicillin induces expression of the beta-lactamase gene in *Staphylococcus*. We now know that penicillin binds to a receptor on the surface of the *Staphylococcus* cells, causing the receptor to become enzymatically active. How does the enzymatically active repressor work?

a. It binds to the promoter of the beta-lactamase gene, stimulating transcription.
b. It binds to the repressor of the beta-lactamase gene and removing it from the operator.
c. It binds to the CRP protein, which then binds the beta-lactamase promoter, thereby activiating transcription.
d. It breaks down the repressor of the beta-lactamase gene.
e. none of the above

691. Understanding how penicillin activates transcription of the beta-lactamase gene in *Staphylococcus* presents opportunities for developing new drugs that will make penicillin effective against these resistant strains. Which of the following hypothetical drugs could prove useful in this regard?

a. a drug that binds to the receptor and prevents binding of the penicillin
b. a drug that binds to the receptor and prevents penicillin from activating the receptor
c. a drug that binds to the receptor and prevents it from breaking down the repressor
d. all of the above
e. none of the above

692. Projects are underway in many research laboratories to genetically engineer mosquitoes to make them more resistant to pathogens that cause human disease. What would such an approach require?

a. a way to induce the mosquitoes to produce greater quantities of defensive proteins that kill pathogens like malaria
b. a way to get the genes that encode the defensive proteins into the mosquito genome (transposons have already been used in this regard)
c. a way to get these genetically engineered mosquitoes to spread into the wild mosquito population
d. all of the above
e. none of the above

693. Edible vaccines involve the genetic engineering of plants so that upon eating the genetically modified plant, the human or animal ingesting the plant becomes vaccinated against a disease-causing organism. Field trials are currently ongoing for which of the following diseases?

a. *Mannheimia haemolytica*, which causes "shipping fever" in cattle. A gene encoding one of the bacterium's proteins was inserted into white clover.
b. *Plasmodium falciparum*, which causes malaria in humans. A gene encoding one of the parasite's proteins was inserted into lettuce.
c. HIV which causes AIDS in humans. An HIV gene was inserted into corn.
d. all of the above
e. none of the above

694. What is the antibiotic of choice for use against *Bacillus anthracis* (the causative agent of the disease anthrax)? Hint: This was the antibiotic that was distributed to postal and other government workers following the bioterrorist attack using *B. anthracis* in Washington DC in 2001.

a. penicillin
b. ampicillin
c. ciprofloxacin
d. erythromycin
e. tetracycline

695. Which human virus has been used as a vector for delivering the normal human CFTR gene to cystic fibrosis patients?

a. Epstein-Barr virus
b. cytomegalovirus
c. adenovirus
d. hepatitis B virus
e. varicella-zoster virus

696. How many molecules of RNA are found in each HIV virus particle?

a. one
b. two
c. three
d. four
e. five

697. Which of the following steps occur in the copying of the (+) strand of RNA of a retrovirus into double-stranded DNA?

a. One of the tRNA molecules carried in the virus hybridizes to one of the (+) ssRNA strands also carried in the virus.
b. Using the tRNA as a primer, reverse transcriptase copies the (+) ssRNA into a (–) ssDNA.
c. Reverse transcriptase degrades the (+) ssRNA, leaving the (–) ssDNA.
d. Reverse transcriptase copies the (–) ssDNA to make the (+) strand of DNA. The DNA is now double-stranded.
e. all of the above
f. only *a*, *b*, and *d*, above

698. HIV is thought to have arisen from which virus by mutation?

a. Rift valley fever virus
b. Crimean-Congo virus
c. SIV
d. Ebola virus
e. Lassa fever virus

699. Once the HIV viral dsDNA has been formed by reverse transcriptase, the HIV dsDNA inserts itself into a chromosome of the host human cells. The enzyme __________________, which is responsible for this insertion, is carried into the human cells within the virion.

a. insertase
b. integrase
c. recombinase
d. protease
e. none of the above

700. The tRNA carried within the HIV virion is crucial to the synthesis of the ds HIV DNA from the single-stranded (+) RNA. What is the source of this tRNA?

a. The tRNA is encoded and synthesized by the host human cell in which the virion was packaged.
b. The tRNA is encoded by the HIV genome, but synthesized by the host human cell machinery in which the virion was packaged.
c. The tRNA is encoded by the host human cell in which the virion is packaged, but is synthesized within the HIV virion.
d. The tRNA is encoded by the HIV genome and is synthesized within the HIV virion.
e. none of the above

701. Consider the following hypothetical gene:

5′-Promotor/Enhancer-Exon1-Intron1-Exon2-Intron2-Exon3-Intron3-Exon4-3′

True or false: After transcription and translation, the amino terminus of the expressed protein will be derived from the 5′-end of the gene.

**702–711. Prenatal Cystic Fibrosis Diagnosis Using Allele-Specific PCR (ASPCR)**

Using the QIAamp DNA Blood Mini Kit (Qiagen) with some minor modifications, DNA is extracted from cell-free maternal plasma (gestational age 11–19.2 weeks) that has been stored at –80 °C after processing. ASPCR amplification was performed on a Gene-Amp PCR system 9700 thermal cycler (Applied Biosystems). Thermal cycling (50 μL) to amplify the 98 bp segment of the D1152H mutation was performed with the following ingredients:

- 8 pmol of forward primer 1152F or Des1152F (identical to 1152F except for the second nucleotide from the 3′ end—A replaces G)
- 8 pmol of reverse primer 1152R (both standard primers designed by Primer Express 1.5 software from Applied Biosystems and synthesized by Proligo)
- the DNA sample
- 1X PCR Buffer with magnesium at a final concentration of 1.5 mM
- dNTPs at a final concentration of 200 μM each from the PCR Core reagent set from Roche
- 1 μL of the Accuprime Taq DNA polymerase from the Accuprime Taq DNA Polymerase System (Invitrogen).

High-accuracy PCR (50 μL) was carried out using

- 30 pmol of forward primer 1152F
- 20 pmol of reverse primer 1152R
- 7.5 μL of 10X Accuprime Pfx Reaction Mix from the Accuprime Pfx DNA Kit (Invitrogen)
- 1 μL of Accuprime Taq DNA polymerase from the Accuprime Taq DNA Polymerase System
- 5X TaqMaster PCR Enhancer (Eppendorf)
- the DNA sample

An assay for the GAPDH gene also was performed by thermocycling, using the forward primer GAPDH-F and the reverse primer GAPDH-R on the ABI Prism 7700 Sequence Detector (Applied Biosystems).

702. The CF mutation for which testing is performed results in what?

  a. the substitution of aspartic acid by histidine
  b. the substitution of a G for a C at nucleotide 3586
  c. the substitution of a G for a C at nucleotide 1152
  d. the substitution of histidine by aspartic acid
  e. Both *a* and *b* are correct.
  f. Both *c* and *d* are correct.

703. True or false: Without any modification, standard ASPCR will differentiate the D1152H mutant allele from the wild-type.

704. Both the standard and high-accuracy assays are designed to do what?

a. detect the presence of a paternal D1152H mutant in the fetal DNA in the presence of wild-type maternal DNA
b. detect the presence of a maternal D1152H mutant in fetal DNA in the presence of other maternal DNA
c. detect the presence of either a paternal or a maternal D1152H mutant in the presence of other maternal DNA
d. detect the presence of free, circulating maternal DNA that contains the D1152H mutant
e. none of the above

705. True or false: The free DNA in plasma represents only fetal DNA.

706. Which of the following might increase the amount of total DNA in a cell-free plasma?

a. pre-eclampsia
b. viral infections that raise the blood cell count
c. delayed processing or mishandling of the blood that results in DNA-containing blood cellular component lysis
d. all of the above
e. only *b* and *c*

707. The high-accuracy assay can detect mutant paternal allele in fetal DNA without a false-positive ______________________________.

a. at a level of five copies or less in the presence of an unlimited amount of maternal wild-type DNA
b. at a level of 300 copies or more in the presence of an unlimited amount of maternal wild-type DNA
c. up to 300 ng of wild-type genomic DNA in the PCR sample
d. only at a level of 300 copies or more in the presence of 300 ng of wild-type genomic DNA in the PCR sample
e. Even the high-accuracy assay has false positives that can be detected using a 3% agarose gel with ethidium bromide staining post-PCR.

708. What is the purpose of the assay for thc GAPDII (glyceraldehyde-3-phosphate dehydrogenase) gene assay?

a. to determine if there is any external contamination in the D1152H PCR assay from previous PCR assays on the same instrument
b. to determine the level of contamination by glyceraldehyde, which is a known inhibitor of PCR
c. to determine the possible presence of unknown inhibitors through the use of a "housekeeping gene" PCR
d. to determine the amount of inhibition of the PCR reaction by the anticoagulant used to keep the whole blood from clotting during the blood draw
e. to determine the total amount of DNA present in the maternal blood sample

709. How do the Accuprime reagents differ from those of other manufacturers?

a. The buffer contains thermostable proteins belonging to the replication accessory protein group.
b. The polymerase is more purified than in other commercial products.
c. Magnesium is replaced with a high level of beryllium, which is a dication that markedly changes the enzymatic properties of Taq polymerase.
d. Magnesium is replaced with a low level of calcium, which is a dication that enhances the "processivity" of Taq polymerase.
e. The dNTPs are replaced with fluorinated analogues.

710. Which of the following is one purpose of the TaqMaster PCR Enhancer in the high-accuracy assay?

a. to increase the "processivity" of the Taq polymerase
b. to reduce primer-dimer artifact formation
c. to enhance annealing between the primers and the mutant allele
d. to disrupt annealing between the primers and the maternal wild-type allele
e. Both *c* and *d* are correct, leading to a greatly favored annealing between the primers and only the mutant allele.

711. The use of the Des1152F primer (which reportedly improves the ASPCR-based allelic differentiation by altering the nucleotide close to the 3′ end of the primer) in place of the perfectly matched 1152F primer ______________________.

a. allows the forward primer to detect mutations other than the D1152H mutation (e.g., ΔF508).
b. allows the forward primer to detect the ΔF508 maternal mutation in the presence of a paternal D1152H mutation.
c. provides some improvement in the specificity of the assay (e.g., better detection of the paternal D1152H allele in the presence of maternal wild-type DNA).
d. improves sensitivity by reducing primer-dimer formation.
e. improves thermocycling assay conditions by raising the temperature required for annealing of primer to the target locus on the DNA.

712. Consider an autosomal dominant inherited disease that demonstrates age-dependent penetrance. At age 40 an individual who believes that he has inherited one allele for the disease from his father remains asymptomatic. The penetrance at age 40 for this disease is 33%. What is the posterior probability that he has inherited the disease, knowing that he is asymptomatic at this age?

a. 50%
b. 43%
c. 40%
d. 10%
e. 100%

713. What does the 5′ end receive before RNA is exported from the nucleus in eukaryotes?

a. an AUG cap so that translation will start every protein (prior to any post-translational modification) with a methionine
b. an aminopterin cap to keep protein from being synthesized in the 5′ direction
c. a poly-A leader to match the poly-A tail, so that the reading frame for translating into protein is bracketed by poly-A
d. a poly-T leader to circularize the mRNA, so that it is not degraded prior to being translated into protein
e. a 7-methylguanosine cap that is linked 5′ to 5′ at the start of the primary transcript

714. Consider an autosomal recessive disease in which the prevalence is 1/2500. Calculate the probability that an individual is a heterozygote or carrier for this disease.

______________________________________________

715. Why is the nucleus of the sperm cell highly condensed and transcriptionally inactive?

a. because only histones H2A and H2B are present
b. because only histones H3 and H4 are present
c. because histones are replaced by fibronectin in the sperm cell
d. because histones are replaced by protamines in the sperm cell
e. because histones are replaced by cumulin in the sperm cell

**716–719.** It was the goal of a study to determine whether the *ABCA3* (ATP-binding cassette transporter A3) gene is involved in surfactant metabolism and whether *ABCA3* is a candidate gene for unexplained surfactant deficiency in full-term infants.

716. Which of the following is/are *true* about the *ABCA3* gene?

a. It is mutated in Tangier disease, which is a disorder involving the accumulation of cholesterol in macrophages and peripheral tissue and a deficiency in high-density lipoproteins.
b. It encodes a 1704-amino acid protein highly expressed in the lung and which has been localized to the limiting membrane of lamellar bodies.
c. It is mutated in several recessive disorders that involve retinal degeneration including macular dystrophy due to Stargardt's disease, most recessive forms of cone-rod dystrophy, and some recessive forms of retinitis pigmentosa.
d. It is expressed in the liver and intestine and mutated in patients with sitosterolemia, a disorder involving the accumulation of cholesterol and other sterols.
e. All of the above are true.

717. *ABCA3* mutations that might lead to clinical surfactant deficiency such as DIP (desquamative interstitial pneumonitis) and/or PAP (pulmonary alveolar proteinosis) are which of the following?

a. W1142X/W1142X
b. G1221S/L982P
c. R106X/R106X
d. L1553P/L1553P
e. all of the above

718. True or false: A variant is more likely to be a deleterious mutation if it is absent in controls and if it affects an amino acid residue that is conserved across several species (e.g., human, mouse, rat, *T. rubripes*, and *D. rerio*).

719. Which of the following mutations in the *ABCA3* gene causes a frameshift regardless of zygosity?

a. G3426A
b. T4657C
c. 4552insT
d. 4909+1G>A
e. T301C

720. Cells activate several genes, not just one, in response to a xenobiotic exposure. The coordinated gene response is accomplished through which of the following?

a. the presence of the same response elements in the promoters of several different genes
b. having many ligand-activated receptors that exhibit a degree of overlap within the ligands that can bind and activate them
c. sharing of response elements rather than ligands
d. all of the above
e. only *b* and *c*

**721–723.** The functional single-nucleotide polymorphisms in the human cytidine deaminase (*HCDA*) gene were investigated in order to test the hypothesis that analytes of drug targets for polymorphism will help to establish gene-based information for the treatment of cancer patients.

721. Which of the following is/are the variant SNP(s) in the human *CDA* gene?

a. A79C
b. G208A
c. T435C
d. all of the above
e. only *a* and *c*

722. Referring to question 721, above, which of the mutations, if any, is silent?

a. A79C
b. G208A
c. T435C
d. both *a* and *c*
e. None of the above. All mutations result in an amino acid change.

723. True or false: The G208A genotype of *HCDA* could possibly contribute in part to slow deaminators.

724. Which of the following statements is/are *true* concerning thiopurine methyltransferase (TPMT)?

a. The enzyme is coded by two separate alleles that yield either the low-activity form, $TPMT^L$, or the high-activity form, $TPMT^H$.
b. The genetic coding for TPMT results in a trimodal distribution of activity (low, intermediate, and high) expressed in 0.3, 11.1, and 88.6% of the population.
c. The expressions of $TPMT^L$ and $TPMT^H$ are 94% and 6%, respectively.
d. The most common of the three polymorphisms for $TPMT^L$ is TPMT*1, which differs from the wild-type (TPMT*3) at residue 165 (G→W).
e. All of the above statements are true.

725. Velocardiofacial syndrome, which is a microdeletion syndrome that demonstrates learning disability/mental retardation, cleft palate, congenital heart defect, characteristic facies, and a predisposition to mental illness, is associated with which of the following chromosomal deletions?

a. 15q11-13
b. 17p11.2
c. 22q11.2
d. 7q11
e. $X_i$

726. If the coefficient of selection, s, is 0.023 for an X-linked recessive disease with an affected frequency ($q^2$) of $5 \times 10^{-4}$, calculate the gene mutation rate per generation ($\mu$).

727. In a microsatellite instability assay for colorectal cancer, which of the following unstable microsatellite markers might be targeted?

a. BAT25 or BAT26
b. D2S123
c. D5S346
d. D17S250
e. all of the above

728. Comparative genomic hybridization (CGH), which is useful for detecting deletions and amplifications in solid tumors, cannot detect which of the following?

a. polyploidy
b. balanced translocations
c. inversions
d. point mutations
e. any of the above

729. The detection of three (3) copies of chromosome X in a metaphase spread might be best accomplished with which of the following techniques?

a. probing with locus-specific, FITC-labeled oligoprobes followed by Sephadex chromatography for removal of excess unbound probe
b. whole chromosome paint
c. nested-PCR
d. nested RT-PCR
e. matrix-assisted laser desorption time-of-flight mass spectrometry

730. A Robertsonian translocation is the result of which of the following?

a. the fusion of the long arms of two acrocentric chromosomes at the centromeres with the loss of both short arms
b. two breaks on a chromosome with inversion of the segment and reinsertion at its original site
c. the breakage of nonhomologous chromosomes with reciprocal exchange of chromosome material
d. the fusion of the short arms of two metacentric chromosomes at the centromeres with the loss of both long arms
e. the fusion of the short arms of two acrocentric chromosomes at the centromeres with the loss of both long arms

**731–734. Detection and Quantification of Mutant Mitochondrial DNA Using a Real-time ARMS qPCR**

731. Which of the following statements concerning mitochondrial DNA (mtDNA) is/are *true*?

a. In the heteroplasmic state, only mutant mtDNA exists within a cell.
b. The most common mutation for respiratory chain defects is A3243G, which is responsible for MELAS (mitochondrial encephalopathy, lactic acidosis, and stroke-like episodes) and MIDD (maternally inherited diabetes and deafness).
c. The proportion of mutant mitochondria is irrelevant to clinical phenotype, as even one mutant mitochondrial genome is sufficient to cause disease.
d. In terms of sensitivity, RFLP-ethidium bromide is superior to radioactive ASO dot-blot analysis.
e. All of the above are true.

732. In the performance of an RT-ARMS-qPCR (real-time amplification refractory mutation system-quantitative PCR) system for A3243G,

________________________________________.

a. the introduction of one, or preferably two, mismatched nucleotides immediately 5′ to the mutation greatly increases the specificity of allele-specific modified primers toward the wild sequence target
b. the introduction of one, or preferably two, mismatched nucleotides immediately 5′ to the mutation greatly increases the specificity of allele-specific modified primers toward the mutant sequence target
c. quantitation is achievable so that the proportion of heteroplasmy can be determined
d. the RT-ARMS-qPCR is found to be superior to TaqMan probes because of nonspecific binding of the mutant TaqMan probe to the wild-type and vice versa
e. all of the above

733. If the threshold cycle ($C_T$) is the cycle at which a significant increase in the reaction product is first detected, and $\Delta C_T$ is $C_T^{\text{wild type}} - C_T^{\text{mutant}}$, the proportion of mutant is which of the following?

a. $1/(1 + 2^{\Delta C_T})$
b. $\Delta C_T/2$
c. $2\ \Delta C_T$
d. $\Delta C_T/(C_T^{\text{wild type}} - C_T^{\text{mutant}})$
e. $\Delta C_T/(C_T^{\text{wild type}})$

734. Which of the following is/are *true* about the RT-ARMS-qPCR assay?

a. It can use total DNA extracted from peripheral lymphocytes using a salting-out method
b. It can employ DNA extracted from muscle by the use of Proteinase K digestion and standard phenol-chloroform extraction and ethanol precipitation.
c. It can be extended to other common mitochondrial mutations such as A8344G for the MERRF (myoclonic epilepsy and ragged red fiber) syndrome; T8993G and T8993C for NARP (neuropathy, ataxia, and retinitis pigmentosa) syndromes; and the detection of homoplasic LHON (Leber herediary optic neuropathy) mutations.
d. It not only detects and quantifies the heteroplasmic mtDNA mutation in a single step, but also makes it possible to determine the total amount of mtDNA genome if a standard nuclear gene target sequence is included in the analysis for normalization.
e. All of the above are true.

735. True or false: According to Haldane's rule, about one-third of all persons affected with a severe X-linked recessive disease will represent new mutations without a prior family history.

**736–741. Acute Myeloid Leukemia (AML)**

It was the goal of two studies to determine the genetic signature of different subclasses of acute myeloid leukemia (AML) using microarray technology.

736. Although there is considerable heterogeneity within risk groups and differences in opinion concerning the significance of a given karyotype in AML, karyotype can be used to risk-stratify patients with AML. Which of the following patients with AML might be considered low-risk?

a. inv(16)
b. t(8;21)
c. t(15;17)
d. t(16;16), which is a variant of inv(16)
e. all of the above

737. Individuals with a normal karyotype could be separated into two subclasses (Group I and Group II). These subclasses showed which of the following general differences (even though both subclasses were similar with respect to age, sex, white cell count, serum lactate dehydrogenase [LD], and presence or absence of an antecedent hematologic disorder)?

a. FLT aberrations were more prevalent in Group I than in Group II.
b. French-American-British (FAB) morphologic subtype M1 or M2 was significantly more common in Group I than in Group II.
c. French-American-British (FAB) morphologic subtype M4 or M5 was more common in Group II than in Group I.
d. Group I is characterized by a high-level expression of GATA2, DNMT3A, and DNMT3B.
e. all of the above

738. For which of the following groups could a genetic signature NOT be determined?

   a. t(15;17)
   b. t(8;21)
   c. 11q23 aberrations
   d. trisomy 8
   e. del(7q)/–7

739. The signature for t(15;17) includes which of the following genes?

   a. those that are associated with abnormalites in hemostasis (*PLAU*, *SERPING1*, *ANXA8*, and *PLAUR*)
   b. those that show resistance to apoptotic stimuli (*TNFRSF4*, *AVEN*, and *BIRC5*)
   c. those that show impairment of retenoic acid-induced cell differentiation (*TBLX1*, *CALR*, and *RARRES3*)
   d. those involved with detoxification of chemical compounds and resistance to chemotherapy (*CYP2E1*, *EPHX1*, *MT1G*, *MT1H*, *MT1L*, *MT2A*, and *MT3*)
   e. all of the above

740. True or false: The t(8;21) signature includes MLT4 (also known as AF6), which is a recurrent fusion partner of MLL leukemias, suggesting a possible shared mechanism contributing to leukogenesis.

741. True or false: High levels of expression of EVI1 in AML occur in approximately 10% of cases and predict a poor outcome.

742. True or false: If a disease gene and its associated marker are nearby, recombination is more likely to occur than if the gene and its marker are well-separated.

743. An important SNP for TNFα, the action of which apparently plays a pivotal role in the inflammatory cascade in Crohn's disease, is which of the following?

   a. a novel silent polymorphism (R196R) in exon 2 of the coding region
   b. a coding polymorphism (M196R or $Met^{196} \rightarrow Arg$) in exon 6
   c. –308 in the promoter region of the *TNF* gene, which produces slightly more TNF than the wild-type
   d. –308 in the promoter region of the *TNF* gene, which produces slightly less TNF than the wild-type
   e. position 308 of the coding region ($C \rightarrow T$ transition), which produces a TNF that is approximately 1.5 times as effective in binding to the receptor site as the normal TNF

744. When loss of fluorescence in a pair of fluorescence resonance energy transfer (FRET) probes occurs due to the removal of a 3′-phosphate group from the acceptor probe, which of the following will attenuate or eliminate such a loss?

   a. minimizing the number of freeze-thaw cycles through which the probes are taken
   b. protecting the probes from light
   c. adding a 3′-C3 carbon spacer to the acceptor probe
   d. Any or all of the above will resolve the loss of 3′-phosphate issue in a FRET analysis.
   e. Only *b* and *c* provide acceptable increased stability with respect to 3′-phosphate loss.

745. True or false: There is an inverse relationship between triplet repeat length and the age of onset in Huntington's disease.

**746–749. A Protocol for the Analysis of the *PML/RAR*-α Fusion Gene in Acute Promyelocytic Leukemia (APL)**

746. The *PML/RAR*-α hybrid gene is found in what percentage of acute promyelocytic leukemia?

a. 40%
b. 50%
c. 75%
d. 95%
e. virtually 100%

747. What does a protocol that encompasses potential breakpoints use?

a. PCR amplification of DNA extracted from mature promyelocytes in peripheral blood
b. RT-PCR using RNA extracted from mature promyelocytes
c. RT-PCR using RNA extracted from reticulocytes
d. a half-nested RT-PCR that employs either an M4 or an M2 primer to encompass bcr 3 or bcr 1/2 (PML) and a reverse primer R5 for the first PCR, and a reverse primer R8 for the second PCR
e. nested PCR using primer M4 and reverse primer R5 for the first PCR and primer M2 and reverse primer M8 for the second PCR

748. True or false: Due to the leucopenic nature of APL, it is highly recommended that bone marrow specimens be analyzed at diagnosis and on followup, with the only exception being for the rare microgranular (variant) M3v.

749. Which of the following controls should be assayed along with the amplified patient DNA?

a. a negative control that is all reagents mixed without sample ("water lane")
b. a positive control, such as the APL derived from the NB4 cell line, or an APL diagnostic RNA with a known PML/RAR-α fusion product
c. a DNA size marker such as ΦX1, a 123-bp ladder, a 100-bp ladder, or a 1-kb ladder
d. all of the above
e. Only *a* and *b* are required for the experiment.

750. Which of the following hemochromatosis (*HFE*) gene combinations is associated with an increased risk of the development of pre-eclampsia in pregnant women?

a. wt/C282Y
b. C282Y/C282Y
c. wt/H63D
d. C282Y/H63D
e. none of the above, as there is no association between the C282Y or H63D allele and an increased risk for the development of pre-eclampsia

ACKNOWLEDGMENTS

1. Structural formulae were produced using the standard version of ChemInnovation Software, Inc., P.O. Box 26561, San Diego, CA 92126.
2. Reagents for the Smith College/New England biologicals seminars were provided by New England Biologicals, Qiagen, and Stratagene.

# Answers and Explanations

1. c.

   Transcription is the DNA-directed synthesis of RNA, translation is the synthesis of protein using an mRNA template, and recombination is the rearrangement of DNA sequences.

2. a.

   Replication is the DNA-directed synthesis of DNA, translation is the synthesis of protein using an mRNA template, and recombination is the rearrangement of DNA sequences.

3. d.

   Prokaryotic DNA is located in the nucleoid region of the cytoplasm. Mitochondria and chloroplasts contain DNA located within these organelles.

4. c.

   Restriction enzymes often recognize DNA sequences with dyad symmetry because they usually bind as a dimer.

5. a.

   Newly synthesized pre-mRNA is processed by the addition of a 5′-cap, a poly-A tail, and the removal of intron sequences followed by the ligation of exons.

6. b.

   Reverse transcriptase uses an RNA template to encode a complementary DNA molecule. T4 DNA ligase will join two strands of DNA by forming a new phosphodiester bond, and DNA polymerase is used in the process of DNA-mediated DNA synthesis (replication). RNA polymerase I transcribes rRNA genes, and RNA polymerase III transcribes primarily tRNA, snRNA, and the 5S rRNA genes.

7. c.

   DNA is typically double stranded, with the strands oriented in antiparallel fashion. RNA is single stranded, but complementary regions can adopt an antiparallel double helical structure; the geometry of such helices is typically A-form.

8. a.

   Eukaryotic DNA consists of multiple linear chromosomes. Prokaryotes often contain several plasmids, but these are small molecules that contain genes encoding for products that are useful to the cell but not essential.

9. b.

Proteinase K is a general protease that is useful in the isolation of genomic DNA because it tolerates high concentrations of detergents used to extract cellular membrane components. In addition it functions at high temperatures to speed up the digestion process.

10. a.

Due to the formation of the fibrin clot, isolation of genomic DNA from clotted blood is very difficult; the DNA isolated is usually of low quantity and poor quality because of mechanical shearing. Heparin is difficult to remove from isolated DNA and can inhibit downstream reactions such as PCR.

11. e.

DNA is soluble in aqueous solutions since it is highly hydrated. Removal of the water of hydration through the addition of salt and alcohol causes the DNA to precipitate out of solution. Removal of the salt and alcohol allows the DNA to again dissolve.

12. a.

DNA absorbs UV light at 260 nm, while protein absorbs at 280 nm due to the presence of aromatic amino acids such as tryptophan and tyrosine. The absolute concentration of DNA may be estimated if the $A_{260}$ is known (a 50 μg/mL solution of genomic DNA has an $A_{260}$ of 1.0). Molecular weight of DNA can be determined by electrophoretic separation through a low percentage agarose gel.

13. d.

All of the other statements are descriptive of <u>plasmid</u> DNA.

14. d.

The 3′-CCA terminus of a mature tRNA molecule is added post-transcriptionally by a separate enzyme. tRNA molecules have double-stranded as well as single-stranded regions; alternative base-pairing such as Hoogsteen base pairs is common in tRNA. A charged tRNA has the amino acid attached to either the 2′ or 3′-OH; the 5′-end of the tRNA is phosphorylated.

15. b.

Messenger RNA contains a tail of up to 200 consecutive A residues; this is added post-transcriptionally during RNA processing. As a consequence of this feature, mRNA can be isolated from total RNA by passing the mixture over a column bound with oligo-dU or oligo-dT; base pairing between the poly-A tail and the oligo will cause it to be retained while the remaining RNA is washed free. The mRNA can then be eluted by disruption of the base pairing.

16. a.

DEPC is used to treat solutions used for RNA work because it inactivates ubiquitous RNases. The DEPC is then removed from the solution by autoclaving. DEPC will destroy amine bases such as TRIS, and so TRIS solutions need to be prepared using pre-treated water.

17. d.

Guanidinium thiocyanate denatures cellular contents and inactivates RNases; phenol-chloroform-isoamyl alcohol is used to extract proteins and protein degradation products. Cesium chloride gradients are often used in conjunction with ultracentrifugation to produce high-purity RNA.

18. c.

Exonucleases are non-specific nucleases that cleave DNA sequentially from the ends, while restriction endonucleases recognize a specific base sequence.

19. c.

DNA ligase will utilize ATP to create a new phosphodiester bond between two DNA fragments where one strand has a 5′-phosphate and the other a 3′-OH.

20. b.

The sequence shown has dyad symmetry. The restriction enzyme will bind as a dimer and cleave both strands between the A and T, yielding blunt-ended digestion products.

21. a.

The sequence shown has dyad symmetry. The restriction enzyme will bind as a dimer and cleave both strands between the two G's, yielding 4-base single stranded overhangs of GATC on each strand.

22. c.

Cleavage at site 3 will produce a 3′-overhang and a recessed 5′-end. As DNA synthesis proceeds in a 5′-3′ direction from an existing 3′-OH, Klenow fragment cannot be used in this instance. Also, Klenow fragment lacks the 3′-5′ exonuclease function possessed by many DNA polymerases. To convert this product to a blunt end, one must remove the 3′-overhang with exonuclease III, which digests single-stranded DNA from the 3′-end. Exonuclease III will continue digestion into the dsDNA molecule, but reaction conditions can be chosen to minimize this, yielding the desired blunt-ended product.

23. d.

Cleavage of both DNAs at site 1 produces complementary ends between the genomic DNA and the plasmid. Hydrogen bonding between these complementary ends holds the fragments in place long enough for DNA ligase to join them together via the formation of a new phosphodiester linkage.

24. c.

Sanger sequencing is also called dideoxy sequencing because it relies on the incorporation of a chain terminating 2′3′-dideoxynucleotide during DNA synthesis. When this nucleotide is incorporated, no further synthesis can occur as there is no 3′-OH for the incoming dNTP to attach to. The Maxam-Gilbert chemical degradation method relies upon specific chemical reactions to cleave DNA after a specific base; reaction conditions are adjusted so that on the average each DNA molecule in the reaction is cleaved once. Edman degradation is a degradation method used to determine the amino acid sequence of proteins.

25. c.

Higher ratios result in too-frequent chain termination and short fragments, while lower ratios do just the opposite, forming longer fragments that are too difficult to resolve on sequencing gels.

26. c.

Keeping the temperature at 94 °C for several minutes ensures that all double-stranded template DNA is denatured to single-stranded. Taq polymerase has optimal activity at 72 °C.

27. d.

PCR primers are usually in the 18–30 nucleotide range; and, depending on their base composition, 50–65 °C is typical for their optimal annealing temperature.

28. d.

New DNA synthesis requires a proper template. Annealing of a primer to ssDNA produces this proper template, with a 3′-OH. GC base pairs have three hydrogen bonds and they are more stable than AT base pairs with only two hydrogen bonds. Annealing of DNA is a two-step process: first, the complementary strands must find each other in solution and start to base pair. Once this has occurred, the remainder of the molecule rapidly zips together into dsDNA. However, if there are many GC pairs near the 3′ end of the primer, the 3′-end of the primer might misanneal to the wrong sequence and be extended before it melts off. This can be avoided by incorporating some A or T residues near the 3′ end of the primer.

29. a.

That a 50 μg/mL DNA solution has an $A_{260}$ of 1.0 holds for genomic DNA, with hundreds or thousands of bases averaging out. With short oligonucleotides this averaging does not occur and a specific formula based on the absorptivities of the individual bases must be used. Commercially synthesized oligonucleotides typically come with a specification sheet that provides the sequence-specific relationship between $A_{260}$ and the concentration for each oligo.

30. a.

The original thermophilic DNA polymerase, *Taq*, was isolated from the bacterium *Thermus aquaticus*. The DNA polymerase gene has since been cloned into *E. coli*. DNA polymerases from other thermophilic bacteria have been used in cycle sequencing and primer extension reactions, but the PCR process specifies a *Taq* polymerase.

31. d.

The ligase chain reaction relies upon the annealing of two adjacent oligonucleotides to each strand of a melted DNA duplex; thermostable DNA ligase will then link the adjacent oligos together. Repeat cycling exponentially amplifies the ligated product.

32. d.

The ligase chain reaction relies upon the ability of DNA ligase to form a phosphodiester bond between two adjacent nucleotides; DNA synthesis is not required.

33. a.

Transcription-mediated amplification and nucleic acid sequence-based amplification are both commercial applications of the original self-sustained sequence replication method. Each method requires RNA polymerase activity as well as RNase H and reverse transcriptase enzymatic activities. The TMA method utilizes a reverse transcriptase with an associated RNase H activity, while the NASBA procedure incorporates the RNase H activity as a separate enzyme.

34. a.

A nucleotide triphosphate (NTP) has the three phosphate groups attached to the 5′-end. The one attached to the ribose through an ester link is the $\alpha$ phosphate. The $\alpha$ phosphate is attached to the $\beta$ via a phosphate anhydride bond, and the $\gamma$ is attached to the $\beta$ via the same linkage. An NTP typically has a free 3′-OH.

35. b.

Strand displacement amplification utilizes restriction enzyme cleavage to create a single-stranded nick in a dsDNA molecule; extension of this nick with DNA polymerase I recreates the restriction site, but the presence of thio-dCTP in the opposite strand prevents its cleavage. The newly synthesized strand is repeatedly extended and cleaved, resulting in exponential amplification of single-stranded product DNA.

36. d.

Theoretically, each cycle of PCR doubles the amount of DNA molecules present in the reaction. Twenty-five cycles will therefore result in $2^{25}$ DNA molecules from each original template DNA molecule. Of course, many factors such as template competition and enzyme degradation serve to limit the amount of DNA actually formed.

37. d.

Gene expression is monitored by measuring the amount of mRNA produced from transcription of the gene of interest. All of the named methods will allow detection and quantitation of the specific mRNA in question.

38. c.

Northern hybridization requires large amounts of RNA and is compromised by sample degradation. Ribonuclease protection assays, while more sensitive, are problematic when analyzing families of related genes, as mismatches between the probe and template provide sites for RNA degradation. RT-PCR avoids these problems, and is capable of theoretically detecting a single copy of template RNA in a sample. However, small differences in amplification efficiency can compromise the strength of signal obtained.

39. d.

In this problem, the upstream and downstream primers are located in separate exons. PCR of genomic DNA will amplify not only the intervening exon sequences, but the intron as well, while RT-PCR will only amplify the exon sequences; the intron was removed during RNA processing. RT-PCR should therefore result in a shorter product.

40. c.

Complementary DNA is DNA that is synthesized using RNA as a template. A, B, and Z DNA are terms that refer to the geometry of the double helix under varying conditions; Z-DNA is a left-handed double helix while A and B are both right-handed.

41. b.

There are several different types of RNases. RNase H will specifically degrade the RNA strand of a RNA:DNA heteroduplex, leaving behind a ssDNA.

42. b.

Random hexamers can be used to prime first-strand synthesis of all RNA molecules in the sample, while gene-specific primers will only prime the specific mRNA in question. All mRNA in the sample can therefore best be primed by using an oligo-dT primer which will hybridize to the poly-A tail found on each mRNA.

43. c.

RNA is a single-stranded molecule, but complementary regions can form double-stranded structures. These structures can interfere with reverse transcriptase "processivity" and so they are removed by melting the RNA at 70 °C first.

44. d.

Each of the primers has significant homology to the other at the 3′-end. As the primers are present in high concentration in the sample, during the annealing step it is likely that one primer will anneal to another primer rather than to a template strand. During extension, each 20-base primer is therefore extended 10 more bases, forming the 30-bp primer dimer artifact. Proper attention during sequence analysis and primer design will avoid these problems.

45. c.

In random sequence DNA, each base has a 1/4 probability of appearing at a given site, and the presence of a base at a particular site has no effect upon the probability of finding a particular base at the next site. If A, T, G, and C are found in equal proportions in the genomic DNA, then the probability of the given hexamer sequence being found is $(1/4)^6$, or 1/4096.

46. d.

Genomic DNA digested by a single six-base cutter results in fragments ranging from a few tens of base pairs to several thousand base pairs; the average of 4096 is just that—an average. To separate long fragments such as these by electrophoresis, the best matrix to use is a low percentage agarose with pore sizes sufficient to allow the large DNA molecules to enter the gel.

47. d.

Lane 4 shows a smear of fragments, representing every possible length of DNA from very long to very short. If the restriction enzyme recognition site is found in a repeated region, then discrete bands may be superimposed on this smear, representing the large amount of specific-sized DNA.

48. a.

λ DNA is cleaved by the restriction enzyme *Hin*d III into discrete fragments with the given sizes. By plotting band migration distance vs. 1/log MW (# of bp), one can estimate the MW of bands seen on an electrophoresis gel.

49. c.

In bacteriophage λ virus DNA, the 23.1 kb fragment and the 4.4 kb fragment lie at opposite termini of the linear molecule. The presence of 12-base-long complementary single-stranded *cos* sites allows the viral DNA to circularize upon *E. coli* infection, linking the two ends of the molecule together. In the restriction-digested sample, these *cos* sites are still present. Heating the sample prior to gel loading will disrupt the base pairing in the *cos* site and will allow the two fragments to migrate separately.

50. c.

Relaxed closed circular plasmid has the largest effective area and therefore will migrate the slowest on the gel. Linear plasmid has the smallest surface area and can snake through the gel pores with relatively greater ease than the open circular form. Sometimes the linear DNA runs a bit slower than the supercoiled DNA.

51. c.

Experience shows that nitrocellulose membranes are somewhat fragile and cannot withstand repeated hybridizations and strippings, while nylon membranes are much stronger.

52. a.

During hybridization, labeled probe hybridizes specifically to the sequence of interest, and nonspecifically to the membrane in general. Nonfat dried milk helps reduce this nonspecific binding by covering these binding sites.

53. d.

As DNA is denatured by heating, the absorbance increases, due to the hypochromic effect. The midpoint of the transition between single-stranded and double-stranded DNA is termed the melting temperature. At this point, one-half of the DNA has denatured and become single stranded.

54. d.

Once DNA has been transferred to a membrane, it must be fixed so that it will not wash off during subsequent hybridization and washing steps. Baking the membrane at 80 °C, or briefly exposing it to UV light, causes covalent bonds to form between the membrane matrix and the DNA. Similar fixation occurs with alkaline transfer of DNA to a negatively charged membrane.

55. e.

Please see the references.

56. a.

Please see the references.

57. c.

Please see the references.

58. f.

Please see the references.

59. b.

Please see the references.

60. h.

Please see the references (keto form shown in the body of the text).

61. i.

Please see the references.

62. g.

Please see the references.

63. d.

Please see the references.

64. j.

Please see the references.

65. b.

Knudson proposed a germ line mutation followed by a later, second, somatic mutation. Other answers may lead to a "double hit," but do not follow Knudson's reasoning.

66. f.

Statements *a*, *b*, and *c* are true because each statement correctly identifies a known form of a zinc finger (please see references).

67. d.

According to Table 11-3 in the reference, each of the cellular proto-oncogenes listed may be amplified in small-cell lung cancer (SCLC). Thus, the correct answer is *d*, which is "All of the above."

68. b.

See page 194 of the Tsongalis and Coleman reference.

69. True.

Because the ethanol was all removed <u>prior</u> to the elution step, this is a true statement.

70. 44 μg/mL.

By simple calculation, 1.1 × 40 ug/mL = 44 ug/mL. Also, please see page 194 of the Tsongalis and Coleman reference.

71. a.

A (Adenine) pairs with T (Thymidine). Thus, poly-A will bind to poly-T presuming that sufficient length for a stable bond exists (*vide infra*), which is the case here.

72. c.

The oligo $(dT)_{25}$ is covalently coupled to a cross-linked cellulose bead. Thus, when the oligo dT recognizes and binds to the poly A tail of the mRNA, the mRNA is essentially also bound to a cellulose particle that can be pelleted with centrifugal force. Please refer to the New England Biolabs reference for more product information.

73. b.

There is no enzymatic activity in the buffer nor is dissolution of the cellulose desirable, since only the mRNA is the required product. The combination of the reduced salt concentration and the elevated temperature is sufficient to break the A–T bonds. Breaking G–C bonds is not required to remove the poly A tail from the poly-dT.

74. True.

Dithiothreitol is added for the purpose of inhibiting RNase.

75. d.

This is the first step in the RT-PCR, which only requires the reverse primer for the synthesis of the first cDNA strand. The temperature is far too low for any denaturation or activation of *Taq* Gold to occur and, thus, the other answers are incorrect.

76. e.

It is here that the first strand of cDNA is synthesized by reverse transcriptase. The temperature is too low for denaturation or activation of *Taq* Gold to occur. Also, annealing of a primer is not required at this stage. Thus, the other answers are incorrect.

77. d.

At this elevated temperature, dissociation of the cDNA-RNA complex, activation of *Taq* Gold, and heat-denaturation of the RT all occur.

78. True.

This statement is true according to the product labeling in the New England Biolabs reference.

79. True.

Although usually not true, this statement is true here because, according to the product labeling in the New England Biolabs reference, each fragment in the ladder is of known DNA quantity and, thus, will yield a fluorescence intensity that is proportional to its concentration after staining with ethidium bromide. By comparing the intensity of fluorescence produced in an unknown with the intensity of fluorescence in a marker band, an estimate of DNA mass in addition to molecular weight can be obtained.

80. a.

The Klenow fragment used in this experiment is a proteolytic product of DNA Polymerase I in which the polymerase activity is retained but the 5′→3′ exonuclease activity is lost and, due to a mutation (D355A, E357A), the 3′→5′ exonuclease activity is abolished. The other answers are incorrect as they are not consistent with the activity of the Klenow fragment.

81. c.

The purpose of using biotinylated random octamers is to provide biotin sites later in the experiment for the well-known, high-avidity biotin-avidin reaction, which will allow detection of the synthesized DNA.

82. True.

The purpose of the LiCl at this point is to precipitate large DNA molecules, but not the small oligonucleotides. This will leave the small oligonucleotides in solution.

83. True.

Formaldehyde is a known denaturant.

84. a.

The transfer of RNA from a gel to a membrane is known as Northern blotting. Southern blotting involves DNA and Western blotting involves protein. Thus, answers *b* and *d* are incorrect. Answer *c* is not a possibility.

85. b.

The purpose of adding formamide to a solution is to cause dsDNA or DNA/RNA duplexes to separate at a lower temperature, not a higher temperature. Formamide has none of the uses described in answers *c* through *e*.

86. True.

The biotin-avidin reaction is one of the strongest, most specific reactions known in biochemistry.

87. True.

As was true in the previous parts of the experiments, it also is true here that an elevated temperature accomplishes, among other things, denaturation of RT.

88. True.

UNG can destroy any carried-over amplicon.

89. d.

At this 50 °C step UNG destroys any unwanted previous product and UNG has optimal activity under the conditions in answer *d*.

90. e.

Exposure to UV light produces both cyclobutane dimers between adjacent thymines and [6-4] products between thymine and cytosine when they are adjacent.

91. True.

This statement is true—see cited reference.

92. e.

See figure 4.1 in the reference, which demonstrates that answer *e* (which states that either *b* or *d* is correct) is the correct choice for this question.

93. d.

*LKB1* (also called *STK11*), which is located on chromosome 19p13.3, is possibly a tumor-suppressor gene, and mutations in or lack of the gene is thought to be the cause of PJS. Also, *LKB1* codes for a serine/threonine kinase, and *LKB1* mutations are found in most, but not all PJS patients.

94. d.

The sensitivity of cells from Fanconi's Anemia patients to the clastogenic (chromosome-breaking) effect of crosslinking agents (e.g., diepoxybutane or DEB) provides an essentially unique cellular marker for diagnosing FA, even in patients who are pre-anemic and who have aplastic anemia or leukemia.

95. c.

The original strand of DNA will be left (unless degraded). The first round of PCR produces a complementary strand that extends from the primer to where the polymerase "falls off" the end of the template strand. Subsequent cycles of PCR produce the desired amplicon defined by only the two primers. However, the longer products will continue to be made each round from the original template strands (so 20 total of each). Thus, for 20 cycles, $2^{20}$ = 1,048,576 copies of the desired amplicon are produced.

96. b.

The 16S rRNA gene is stated in Table 10.24 of the reference.

97. a.

This is the stated purpose for amplicon containment in Table 10.24 of the reference.

98. b.

This is a stated requirement in Table 10.24 of the cited reference.

99. True.

Some versions, such as those for the COBAS, and the newest versions of the MWP assays (e.g., MTB), contain an internal control.

100. d.

With allele-specific hybridization, several separate specimens can be screened simultaneously for a single sequence of interest by probing with a labeled oligonucleotide. The process is very specific, because it relies upon subtle differences in duplex stability caused by even a single mismatched base pair.

101. b.

This process is the opposite of ASO hybridization. In this procedure, a labeled probe is prepared by PCR and used to hybridize a membrane. Several different oligonucleotides are spotted onto this membrane. Only one specimen at a time can be screened, but one can screen for many different potential mutations during that screening.

102. b.

Neuropathy induced by acrylamide is a toxic distal axonopathy that begins with degeneration of the nerve terminal. Although most human poisonings have been largely limited to those that occur in factory and construction workers who have been exposed to high doses of acrylamide, cautious handling of acrylamide in the laboratory is advised. Acrylamide monomers, both acrylamide and bis-acrylamide, are neurotoxins; care must be taken when working with such solutions to avoid adverse exposure. Once the solutions have polymerized, however, they are safe to handle.

103. a.

Formamide is another denaturant that is often used; denaturing agarose gels are run with 0.4 M NaOH. Each of these solutions disrupts hydrogen bonding between base pairs and allows the strands to separate.

104. a.

SSCP analysis is based on the fact that a single-stranded DNA, much like RNA, will attempt to fold as complementary regions interact together to form regions of double-stranded structure. This folding is sequence specific and imparts a unique conformation to the strand. SSCP gels must be run under native conditions so as to preserve this specific conformation to permit separation of the two strands by electrophoresis. Gels must be run at cold temperatures as even a small amount of joule heating can be enough to melt the short-double stranded regions.

105. c.

The procedure relies on subtle changes in shape imparted by a single base pair mismatch in the DNA; such a change in shape causes the duplex to migrate slightly slower through the gel, permitting separation. A similar effect is observed when genomic DNA that is heterozygous for a particular mutation is amplified, as amplified normal allele strands can form duplexes with mutant allele strands during the PCR process.

106. b.

Melting of DNA is a gradual process, with less stable DNA melting at lower temperatures than more stable DNA. Once the DNA has melted in a short region, the conformation of the molecule changes, retarding significantly its progress through the gel. Often a highly stable region of ~40 GC base pairs is incorporated into the 5′ or 3′ ends of the DNA during PCR of the region of interest; this is called a GC-clamp. This provides a highly stable region that will melt last, allowing very subtle melting temperature differences (due to mutations) to become much more apparent.

107. d.

Primer B is identical to Primer A except for the base at the 3′ end. Improper annealing of this base to the template strand precludes any extension from the 3′-OH.

108. c.

Patient F shows nine separate bands, indicating that all regions of interest are present in the genomic DNA and can thus be amplified. Patients D, E, and G, on the other hand, have had one or more exon deletions, with the result that primer binding sites for these PCRs are not present.

109. b.

Patient E has had three exons deleted; as a result, primer binding sites for these PCRs are not present.

110. c.

Exonuclease III is an enzyme that degrades DNA in a 3′–5′ direction; it is active on single-stranded DNA, and to a lesser extent on double-stranded DNA. In multiplex PCR, many separate amplifications are going on under the same conditions; it is therefore necessary to ensure that sufficient dNTPs are present to handle all of the reactions.

111. d.

Dr. E. M. Southern first published the DNA blotting process in 1975. This process revolutionized molecular biology, because it became possible to study a single specific DNA sequence against a background of other sequences. Northern blotting is an analogous procedure in which RNA is transferred to a membrane by blotting.

112. b.

Southern blotting is the actual transfer procedure by which DNA is transferred from an agarose gel to a membrane. Once transferred, it must still be probed with a labeled probe for the sequence of interest.

113. a.

Ethidium bromide is a flat, planar hydrophobic molecule that intercalates between base pairs of nucleic acid. It is fluorescent while free in solution, but its fluorescence increases dramatically when bound to DNA. This is because, while intercalated, it is relatively shielded from quenching interactions with the solvent.

114. b.

Ethidium bromide binds DNA by intercalation between base pairs. Double-stranded DNA has many more base pairs than does single-stranded DNA that has folded somewhat into short double-stranded regions.

Note from author Steven Williams: Although *b* is the standard accepted answer, perhaps something else is going on. For example, ethidium bromide binds very well to RNA that has been run on a denaturing gel. In fact, it binds so well to the ssRNA that one can add it to the RNA and load it directly into the well with no EtBr in the gel or buffer and obtain great results with no post-staining required. One cannot accomplish the same thing with DNA.

115. a.

Capillary transfer of nucleic acid from gel to membrane is very simple and is often performed with such mundane equipment as Pyrex baking dishes and paper towels. Vacuum transfer and electroblotting, while somewhat faster at transferring nucleic acids, require special equipment.

116. Electroblotting has the highest efficiency of transfer. This method would most likely lead to a successful analysis.

117. a.

DNA that is to be transferred from a gel to a membrane must be single stranded. This is usually accomplished by soaking the gel in a solution of NaOH following electrophoresis. It is also possible to run the gel under denaturing conditions by using NaOH as the running buffer; in this case the DNA is ready to transfer immediately after electrophoresis is completed.

118. a.

The shotgun sequencing approach builds up a master sequence directly from the sequencing of short, individual fragments. The final sequence is assembled by using overlapping regions of the short fragments to assemble longer sequences called contigs.

119. b.

Increased vacuum strength can collapse the gel, making it effectively a higher percentage with smaller pores and thus retarding blotting speed.

120. b.

Transferred DNA is already denatured and hybridization with an unlabeled probe would be pointless, as it would be impossible to locate the sequence of interest.

121. b.

RNA, unlike DNA, is labile in basic solutions due to the presence of the 2′-OH group on the ribose sugar. Electrophoresis of RNA solutions is best accomplished at neutral pH.

122. d.

Formaldehyde and glyoxal both disrupt hydrogen bonding and permit the RNA to migrate through gel in pure single-stranded form.

123. b.

Solutions used for RNA work are best kept at near-neutral pH.

124. c.

When setting up a Southern or Northern blot, care must be taken to remove all bubbles between the gel and the membrane. The presence of an unnoticed bubble has ruined many critical experiments by placing air between the gel and the transfer membrane, negating transfer in that area.

125. a.

Treatment of *E. coli* DNA polymerase I with the protease Subtilisin cleaves the molecule into two pieces. The small fragment contains the 5′–3′ exonuclease activity, while the large fragment retains the DNA polymerase and 3′–5′ exonuclease activities.

126. b.

DNA polymerase I will extend a single strand break in DNA via its 5′–3′ exonuclease and polymerase activities. If labeled dNTPs are incorporated into the reaction mixture, then the newly formed DNA will bear the label.

127. c.

The Klenow fragment is used in this reaction, since it does not contain a 5′-3′ exonuclease activity. If it did, the extension of one primer could result in the degradation of DNA downstream that has already been synthesized, resulting in dramatically reduced labeling efficiency.

128. b.

Polynucleotide kinase will transfer the labeled γ-phosphate of ATP to the 5′-OH group of a DNA strand. The enzyme from T4 phage is most frequently used for this purpose.

129. c.

Nucleic acid strands that are complementary will form double-stranded regions. Riboprobes are single-stranded RNA probes. An RNA:RNA duplex, due to the geometry of the helix, is stronger than an RNA:DNA duplex or a DNA:DNA duplex with the same sequence.

130. b.

Due to the inherent instability of RNA, heat and alkaline solutions will efficiently remove a labeled riboprobe. Following sufficient washings the membrane can then be hybridized with a different probe.

131. c.

Some satellite probes hybridize to chromosome-specific, repetitive DNA sequences located near the centromeres of chromosomes in metaphase or interphase nuclei. The number of signals observed can be useful in the detection of aneuploidy. Unique sequence probes hybridize to a specific gene in the chromosome, and chromosome paints contain chromosome-specific DNA libraries that hybridize to sequences throughout a specific chromosome.

132. b.

*In situ* PCR is much more inefficient than solution PCR, with only several-fold amplification observed.

133. a.

When human chromosomal DNA is broken into fragments 50–100 kb in length and is fractionated by density gradient centrifugation, the majority of the DNA localizes to a single band. Three satellite bands of differing densities are also observed; these bands contain DNA that is highly repetitive and has GC contents that differ from the genome average (40.3%). The repeat unit can be any length from $< 5$ bp to $> 200$ bp.

134. b.

Minisatellites form clusters of up to 20 kb in length; the repeat unit typically contains fewer than 25 bp.

135. c.

Microsatellites consist of short repeat units, 2–4 bp in length, and repeated only 10–20 times. These sequences are also termed STRs, or short tandem repeats.

136. b.

EcoRI is a "six-cutter" that cuts between G and A in a 5′-GAATTC-3′/3′-CTTAAG-5′ sequence segment and leaves G paired with C and 5′-AATT overhangs on both ends.

137. a.

Even though the optimal temperature is 37 °C, the enzyme is heat-inactivated at 65 °C and some heat-inactivation is anticipated at the lower incubation temperature over an extended period of time (overnight).

138. False.

The knowledge that this statement is false comes from practical experience. The low-melt gels are fragile and very slippery. Handling low melt gels is difficult, and even re-positioning them in the electrophoresis chamber in which they were cast can cause them to break.

139. True.

Again, the truth of this statement comes from practical experience. Ficoll is added to an electrophoresis/marker mixture so that the entire mixture will drop to the bottom of the electrophoresis well, where it can be pulled into the gel by electric current.

140. c.

From this point in the experiment it is undesirable to have enzyme that will cut product. Please see question 137., above, for the heat stability/lability of EcoRI.

141. a.

Please see the product labeling found in the New England Biolabs reference.

142. a.

UV light does not affect whether DNA is double stranded or single stranded, nor does it affect the annealing processes. However, UV light can cause the formation of cyclobutane dimers between adjacent thymines and [6-4] products between thymine and cytosine when they are adjacent. Economics (answer *e*) may be an issue but it isn't a reason here. The long-wavelength UV lamps are not necessarily cheaper.

143. e.

The temperature is not high enough to melt the DNA into single strands, but it is high enough to melt the low-melt agarose.

144. b.

The enzyme beta-agarase will function at 41 °C without significant denaturation, while the agarose will not significantly re-solidify.

145. a.

The purpose of the agarase digestion was to cleave the agarose into smaller fragments (saccharides), leaving the DNA intact. Thus, the correct answer is *a* since the smaller particles should pass through the filter and retain the larger molecules. (DNA should be retained by the filter that is designed to retain >125 bp dsDNA and >300 nucleotides for ssDNA.)

146. False.

While it is true that the λZAP Express Vector has only one EcoRI site, the wild type has five EcoRI sites. Thus, the statement is false.

147. True.

The statement is true since the DNA must be packaged in lambda head and tail fibers in order to be infectious to the *E. coli.*

148. True.

Linear DNA in *E. coli* is quickly degraded while circular DNA is not.

149. True.

A sufficient number of cells must remain uninfected in order for the solid, smooth lawn of *E. coli* cells to cover the plate.

150. $1.325 \times 10^5$.

(Plaques/2) × 10 (dilution factor) = pfu/μL × 530 μL (total volume) = pfu in total library tube.

151. c.

A clear spot indicates that *E. coli* were lysed.

152. c.

The NaOH is required to break open the phage and release the DNA, which is converted from ds to ss by the NaOH. If exposure to NaOH is continued, the nylon will be hydrolyzed completely to monomers.

153. d.

Although some nucleotide dimers may be formed, the main purpose is to fix (chemically bond) the DNA to the exposed -$COO^-$ and $H_2N$-groups on the nylon membrane.

154. False.

The first (and, to a lesser extent, second and following) plaque lift does not capture all of the bacteriophage lambda DNA, allowing for additional plaque lifts.

155. False.

Unlike nitrocellulose, which will autoignite in the presence of oxygen at elevated temperatures, nylon is relatively stable in air at 80 °C.

156. a.

The addition of salmon sperm DNA causes the large excess of the salmon sperm DNA to cover spots on the fixed membrane that will give rise to non-specific binding. Thus, the non-specific binding of the DNA probe is eliminated by allowing non-specific binding of an unlabeled DNA to occur prior to the addition of the DNA probe.

157. e.

The elevated temperature is sufficient to melt dsDNA and activate Taq Gold. The temperature is far too high for routine dissolution of reaction components.

158. True.

The final 72 °C for 7 minutes allows for any final extension by the polymerase.

159. e.

The temperature is sufficient to melt dsDNA. The temperature is not sufficient to cause electronic excitation of the fluorescein or break a common chemical bond. The elevated temperature also has nothing to do with activating or deactivating the stable enzyme HRP.

160. a.

The secondary purpose of the 94 °C step is to lyse any remaining lambda phage and denature the lambda DNA from double stranded to single stranded.

161. True.

The statement is true. By simple deduction the dissociation (melting)/re-association temperature of primer/template DNA complexes for both primers needs to be very close, so that both primer-template complexes will be associated for extension or dissociated at the same time in the PCR.

162. c.

Only the long, basic DNA strands will be retained while the smaller oligos and dNTPs are eluted.

163. True.

Lowering the pH makes the DNA more acidic and more likely to bind to the silica.

164. e.

The large band or “blob” (low-molecular-weight material) is indeed unincorporated FL-dUTP (i.e., FL-dUTP that was not incorporated in the labeling of the 1.4 kb probe).

165. False.

By observation, one notices that sufficient fluorescence and, therefore, ethidium bromide, still exists to make the gel toxic and mutagenic. These gels should be disposed of in a manner consistent with your organization's safety rules.

166. e.

All of the statements concerning plasmids and $\lambda$ are true.

167. a.

The "A" overhang does indeed allow for the PCR product to readily ligate to the T overhangs of the cut plasmid.

168. True.

Even though the *E. coli* cells were grown specially to be competent cells, they still must be heat-shocked in order to take up the DNA.

169. True.

As was stated earlier for another experiment, normally this statement would be false. However, the lambda DNA ladder used in this experiment has bands not only of known molecular weight, but also of known concentration. Thus, an estimate of a band's mass may be obtained by qualitative comparison to one of the marker bands.

170. b.

The large, lower-molecular-weight band farthest down is the supercoiled form of the plasmid.

171. d.

The peaks below approximately 50 and those above approximately 550 are too broad and coalesced to be valuable for analysis.

172. b.

Even though xeroderma pigmentosum (XP), Cockayne syndrome (CS), and the photosensitive form of trichothiodystropy (TTD) are all rare autosomal recessive syndromes associated with a nucleotide excision repair (NER) defect, only XP patients also exhibit a greater than 1000-fold increased frequency of sunlight-induced cancers.

173. b.

See the cited reference.

174. True.

See "Imprinting" in the cited reference.

175. e.

See "Molecular Biology of *p53*" in the cited reference.

176. True.

See Table 23-1 in the cited reference.

177. d.

See "Molecular Biology of *p53*" in the cited reference.

178. e.

See "Molecular Biology of *p53*" in the cited reference.

179. d.

See "The Sex Chromosomes" in the cited reference.

180. d.

The correct is *d*, or "All of the above," even though hPMS1 appears to be rare. Please see Table 18-7 in the reference.

181. b.

See Figure 18-3 in the reference.

182. a.

See "The Diagnosis and Management of HNPCC" in the reference.

183. e.

See "Q and G bands" in the reference.

184. a.

See "tRNA Molecules Match Amino Acids to Codons in mRNA" in the reference.

185. d.

See reference, under "The Genetics of Retinoblastoma."

186. True.

See cited reference.

187. True.

See reference, under "Molecular Genetics."

188. e.

All statements are correct, according to "Biochemical Characterization of the Retinoblastoma Gene Product" in the cited reference.

189. d.

See "Involvement of pp110$^{RB}$ in the Cell Cycle" in the cited reference.

190. e.

According to the reference, any of the pathways is possible.

191. True.

See "Involvement of pp110$^{RB}$ in the Cell Cycle" in the cited reference.

192. a.

See the reference, specifically "Genetic Aspects of NF2."

193. c.

See "Genetic Aspects of NF2" in the cited reference.

194. c.

See the introduction in the cited reference.

195. a.

See "Genetic Aspects of NF2" in the cited reference.

196. False.

The *NF1* gene resides on chromosome 17q11.2 and the gene for *NF2* is on chromosome 22q.

197. c.

See "Silencing the V*HL* Gene by DNA Methylation" in the cited reference.

198. b.

Of the diseases listed, only Leber's hereditary optic neuropathy results from mitochondrial inheritance. Cystic fibrosis is autosomal recessive (chromosome 7p31.2); Fragile X is X-linked dominant; Huntington's disease is autosomal dominant (4p16.3); and Charcot-Marie-Tooth disease also is autosomal dominant (17p11.2, Type 1A neuropathy).

199. c.

See Table 5.2 in the reference.

200. a.

See Figure 2-3 in the reference.

201. b.

202. c.

203. f.

204. e.

205. p.

206. a.

207. h.

208. j.

209. i.

210. k.

211. d.

The correct answers for 201—211 are from Tables 1.1 and 1.2 in the reference.

212. e.

According to the reference (see "Polymerase Chain Reaction, In-House [Home-Brew] Assays, Technical Considerations"), all of the statements are correct.

213. d.

RNases are ubiquitous and are a frequent cause of failure in assays. Separation of serum or plasma from cells immediately and storage at –20 ° C reduces these problems.

214. a.

A significant problem often encountered in the isolation of RNA is its degradation by RNases. Disruption of cells releases these RNases, which can compromise an experiment even before it gets started. Disruption of the cells with a chaotropic agent such as GTC accomplishes two feats simultaneously: (1) release of cellular RNA and (2) denaturation of RNases.

215. c.

DEPC is a highly active alkylating agent that destroys the enzymatic activity of RNases by reacting with histidine residues in the protein. It also reacts with amine buffers such as TRIS, and so these solutions must be made up with water that has been pre-treated with DEPC and then autoclaved to remove the last traces of DEPC prior to buffer preparation.

216. a.

Nucleic acids are soluble because they contain several waters of hydration. Dehydration of the molecule through the addition of salt and alcohol precipitates the nucleic acid.

217. c.

Complementary DNA is a strand of DNA that is produced by the action of the enzyme reverse transcriptase, using RNA as a template.

218. d.

The correct answer is *d* because DNA polymerases utilize DNA as a template to direct the synthesis of a new DNA strand; RT uses an RNA strand as a template.

219. d.

By convention, nucleic acid sequences are read in a 5′–3′ direction, from left to right unless otherwise specified. Complementarity between strands dictates that G pairs with C and A pairs with U.

220. b.

Nucleotide triphosphates are used in the synthesis of RNA; to synthesize deoxyribonucleic acid, one must use a dNTP.

221. b.

Random octamers anneal to any complementary sequence to prime synthesis, while oligo-dT anneals to the poly-A tail of eukaryotic mRNA. The 13-base oligonucleotide will anneal only to a particular sequence and is therefore most specific for HCV.

222. c.

Despite the most careful protocols and technique, RNases very frequently contaminate a reaction somewhere along the way. RNasin is a commercial product that inhibits the action of RNases.

223. b.

When setting up any molecular biological reaction, the enzyme is typically added last. This ensures that the enzyme remains at its stable, cold temperature until it is necessary to begin the reaction. In PCR, it also helps to prevent contamination of *Taq* polymerase with template DNA.

224. c.

1.0 μM is equivalent to 1.0 pmol/μL. The reaction therefore contains 100 pmol of primer and thus requires 10.00 μL of primer stock solution.

225. b.

When preparing a master mix for use in setting up several related reactions, components that are common to each reaction can be included in a master mix, to avoid errors when pipetting very small sample volumes.

226. d.

Incorporation of dUTP, followed by treatment with uracil-N-glycosylase, will degrade PCR products from positive reactions and will prevent carryover into subsequent reactions.

227. b.

PCR products are typically a few hundred base pairs in length and are best characterized on a moderate percentage agarose gel; 15% acrylamide gels have extremely small pore sizes and are suitable for the purification of short pieces of DNA.

228. b.

The presence of an amplified product of the expected size is a positive reaction. Assuming the appropriate positive and negative controls gave the expected values, this result indicates HCV infection.

229. a.

RNA from a stock strain of HCV, included in a parallel specimen tested with the unknowns, will provide control that each step of the assay functioned properly and that no RNases were introduced, resulting in a false negative report.

230. b.

Transcription is the process by which DNA-dependent RNA polymerase synthesizes an RNA molecule from a gene.

231. a.

Ethidium bromide is a fluorescent molecule that is commonly used to detect double-stranded nucleic acids.

232. d.

Most eukaryotic genes have their coding sequence broken into exons interrupted by noncoding intron sequence. These introns are spliced out during the processing, which leads to the formation of mature mRNA.

233. e.

Single-stranded nucleic acids, with their exposed hydrophobic bases, are inherently unstable in aqueous solution, and will seek to anneal to their complementary sequence to form a double-stranded structure with the bases located interior to the molecule and shielded from solvent.

234. f.

RNA is alkali labile. A buffer at neutral pH is frequently used in RNA electrophoresis.

235. c.

B-form DNA is the "normal" form of DNA found in solution. It represents an antiparallel right-handed double helix with 10 bp/turn. Other forms of DNA are A-form and Z-form.

236. i.

The Klenow fragment is the large fragment of *E. coli* DNA polymerase I that has been proteolytically treated to remove that portion of the enzyme containing the 5′–3′ exonuclease activity.

237. g.

Sequenase is the trade name for a genetically modified form of T7 DNA polymerase; as its name suggests, it is used in DNA sequencing reactions.

238. j.

Many plasmid vectors contain promoters for RNA polymerases on either side of a multiple cloning site, allowing for in vitro production of strand-specific RNA from a cloned DNA fragment.

239. f.

Most endonucleases cleave internal phosphodiester bonds to leave a 5′-phosphate and 3′-OH; micrococcal nuclease is atypical in producing the opposite types of fragments.

240. d.

Retroviruses are RNA viruses that utilize the enzyme reverse transcriptase to produce a DNA copy of their genome prior to integrating into the host.

241. e.

DNase I is an endonuclease that either creates single-stranded ($Mn^{2+}$) or double-stranded ($Mg^{2+}$) breaks in DNA, depending on the metal cation present in the reaction mixture.

242. c.

DNA ligase catalyzes the formation of a phosphodiester bond between a 3′-OH and a 5′-phosphate; ATP is required for the reaction. The template can be either blunt ended or can have staggered ends.

243. b.

Uracil is not normally found in DNA, but is found in RNA; in DNA the base thymine is used instead. Uracil is formed from the spontaneous deamination of cytosine residues; if this damage were not corrected then a GC base pair will be converted to an AT base pair following replication.

244. a.

The initial PCR method utilized Klenow DNA polymerase, and a fresh aliquot of enzyme had to be added after each step. Modern implementation of the method requires the presence of a thermostable DNA polymerase (such as *Taq* polymerase) for optimum results.

245. h.

DNA ends produced by nuclease digestion typically contain a 5′-phosphate group. This group must be removed if the DNA fragment is to be end-labeled with $^{32}P$ by polynucleotide kinase. In very high concentrations alkaline phosphatase will also remove a 3′-phosphate group.

246. c.

NASBA is a proprietary implementation of the SSSR amplification technique. NASBA is an isothermal RNA amplification method using reverse transcriptase to create a cDNA from an RNA molecule, and RNase H to degrade the original RNA strand prior to second-strand DNA synthesis.

247. d.

Ligase Chain Reaction (LCR) only requires the presence of adjacently annealed oligonucleotides and a thermostable DNA ligase. A variant procedure with higher specificity (called gapped LCR) requires the presence of dNTPs and a DNA polymerase to extend one oligonucleotide until it becomes adjacent to the neighboring oligo; typically this is only a distance of one or a few bases.

248. c.

There are two primary categories of amplification methods: (1) target and (2) signal amplification. In signal amplification methods, a few molecules give rise to a very large signal, usually mediated by enzymatic reactions. Branched DNA is a signal amplification method. In target amplification methods, a few target molecules are amplified to a very large number of target molecules; each one gives a small signal, but, in aggregate, the signal produced is very large.

249. c.

Strand-specific probes are useful in the investigation of, for example, strand-specific DNA repair of actively expressed genes.

250. d.

Nick translation incorporates dNTPs into DNA, not NTPs. The label can either be an $\alpha$-[$^{32}$P]-dNTP or the label can be nonradioactive and attached to the base by means of a long spacer arm.

251. a.

T4 PNK catalyzes the transfer of the $\gamma$-phosphate group of ATP to the 5′-OH group of a DNA molecule.

The correct answers for questions 252–261 are in Table 11-2 in the listed reference.

252. j.

253. d.

254. k.

255. c.

256. l.

257. i.

258. g.

259. m.

260. e.

261. a.

262. True.

The isolated mRNA has already been processed and is ready to be transcribed.

263. a.

The purpose of the DEPC (and the autoclaving of an RNase inhibitor) is to ensure that no RNase activity will be present.

264. c.

Binding of the linker-primer to the mRNA occurs *via* the poly-A tail annealing to the $(T)_{13}$ portion of the linker-primer.

265. False.

The reverse transcriptase (RT) has no demethylase activity.

266. c.

Even though the Polymerase I keeps on synthesizing to the 5′ end of the template strand, about 5–15 (5′-UTR or untranslated region) nucleotides are lost at the 3′ end of the template strand of DNA.

267. True.

The purpose of the blunting mixture is to remove the 5–15 nucleotides from the 3′ overhang on the first strand of cDNA.

268. b.

After the addition of both linkers, an AATT overhang exists on both strands of DNA at the 5′ ends.

269. False.

The T4 polynucleotide kinase phosphorylates only the 5′ ends.

270. d.

Heating to 70 °C is a simple method for inactivating many enzymes.

271. c.

XhoI is a "six cutter" that cuts between C and T in a CTCGAG/GAGCTC sequence only at the end of the cDNA that represents the poly-A tail of the original mRNA.

272. True.

Any XhoI sites within the cDNA (excluding the linker-primer region) are methylated on the first strand cDNA. Such half-(hemi)methylation is sufficient to block cutting by XhoI.

273. e.

The purpose of the Chromaspin column is to allow the ds cDNA to pass through while retaining small fragments of cDNA and excess adaptors from the synthetic reactions.

274. d.

The mixture of chloroform and isoamyl alcohol extracts proteins, degraded proteins, and phenol from the aqueous phase that contains the DNA.

275. a.

There is enough phenol and $CHCl_3$ present to make the less dense layer the upper layer after centrifugation.

276. c.

The purpose of the diffusion using a membrane that only allows the passage of small molecules is to move small molecules and ions, such as salts, from the area of higher concentration (droplet on the membrane) into the area of lower concentration (water below the membrane). This dialysis will also remove individual free nucleotides and amino acids.

277. a.

An EcoRI site was created on one end while an XhoI site was created on the other end of the inserted cDNA. Thus, the insert can only go in such that an EcoRI end matches an EcoRI end and an XhoI end matches an XhoI end.

278. False.

Freezing this solution will destroy the phage.

279. True.

There will be cell lysis but not formation of blue color (5-bromo-4-chloro-indigo) in the plaques that contain no insert because these phage lack an intact gene for β-galactosidase.

280. True.

When there is an intact *lac*Z gene, the enzyme (β-galactosidase), which is capable of hydrolyzing the indolyl-galactoside bond, is produced, and the blue dye will be one product of the reaction.

281. b.

See Figure 15-1 in the reference.

The correct answers to questions 282–289 are given as definitions in the references.

282. h.

283. e.

284. f.

285. a.

286. c.

287. d.

288. i.

289. g.

290. a.

See "Molecular Genetics of BLM" in the listed reference.

291. c.

See "Clinical Aspects of Breast Cancer" in the listed reference.

292. e.

Any of the techniques may be used for the analysis of *BRCA1* mutations (see the reference, under "*BRCA1*").

293. True.

See "*BRCA1*" in the listed reference.

294. True.

See "*BRCA1*" in the listed reference.

295. e.

See the reference.

296. True.

See "*BRCA2*" in the listed reference.

297. a.

See "The Role of HPV Infection in Cervical Tumorigenesis" in the listed reference.

298. c.

See "The Role of HPV Infection in Cervical Tumorigenesis" in the listed reference.

299. d.

See "The Role of HPV Infection in Cervical Tumorigenesis" in the listed reference.

300. True.

See "WRN Mutations" in the listed reference.

301. False.

Hutchinson-Gilford patients do not have WRN mutations (see "WRN Mutations" in the listed reference).

302. c.

See "*In Vitro* WRNp Studies" in the listed reference.

303. d.

However, some bacteria also use GUG or UUG as the start codon, as well as the typical AUG.

304. a.

The genetic code, with but very few exceptions, is universal; the few changes from this paradigm typically involve the termination codons. A phylogeny of the changes seen in mitochondria suggests that the universal code was changed at several points during mitochondrial evolution. Since mitochondrial genomes code for only ~10 proteins, the problem of disruption in protein synthesis caused by changes in the genetic code is much less severe than if it were to occur in the nuclear genome.

305. b.

A nonsense mutation is one in which a change in base sequence changes the codon from one coding for an amino acid to a "Stop" signal. A suppressor is a mutation that compensates the effects of a primary mutation, so that the combination of the two mutations restores wild type phenotype. In the case of a nonsense mutation, the suppressor occurs in a tRNA gene, allowing the tRNA to recognize a stop codon and insert an amino acid, allowing translation to continue.

306. d.

Following their synthesis, the default location for proteins is to remain in the cytosol. If a protein is to move to another compartment of the cell, such as the mitochondrion, a specific signal must be included in the peptide chain. The type of signal directs where the protein ends up.

307. d.

Signals targeting a protein to the mitochondrion or chloroplast are located in the N-terminal region of the protein. Nuclear localization signals are internal, and peroxisomal signals are in the C-terminus.

308. c.

The process of protein folding to an active form can be spontaneous (such as in the case of ribonuclease A). However, some proteins cannot find the proper conformation on their own and need the assistance of helper proteins, called chaperones, in order to avoid folding into potentially stable but inactive structures.

309. d.

TOM and TIM are acronyms that stand for "Transport across the Outer or Inner Membrane" of the mitochondrion, so that the translocated protein ends up inside the mitochondrial matrix.

310. d.

The nuclear localization signal is a 7–9 amino acid stretch of basic or bipartite amino acids.

311. b.

Cytosolic protein degradation occurs as a large complex called the proteasome that recognizes and binds to a targeted protein. The targeting consists in the covalent attachment of the small protein, Ubiquitin, to the target protein via an isopeptide bond between the C-terminal Glycine of Ubiquitin and a γ-amino group of Lysine in the target protein. Polyubiquitination can occur using $Lys^{46}$ of Ubiquitin as a target site for further attachment.

312. b.

There are three eukaryotic RNA polymerases: I transcribes most rRNA genes, III transcribes the 5S rRNA gene and tRNA genes, while II transcribes mRNA genes.

313. a.

The consensus prokaryotic promoter contains a highly AT-rich sequence TATAAT located approximately 10 bases upstream of the start point. In eukaryotes a similar sequence called the TATA-box is found about 35 bases upstream of the start point.

314. c.

As RNA polymerase moves along a double-stranded template DNA strand, a region of DNA is melted to form a "bubble," where the polymerase can gain access to the DNA to direct the synthesis of a complementary RNA molecule. The length of this transcription bubble varies in length from 12–20 base pairs, depending on the phase of the elongation cycle and the methods used to map it. At any given time the most recently synthesized approximately 25 bases of RNA are still complexed with DNA and/or the enzyme; the remainder of the synthesized RNA has unwound from the DNA and has emerged from the back of the polymerase.

315. d.

DNA replication occurs at a rate of ~800 bp/second, but the rate of transcription is much slower and corresponds approximately to the rate of translation (~15 amino acids/second).

316. a.

When a translocating RNA polymerase pauses for some reason (perhaps it has encountered a damaged base in the template strand), a defined sequence of events is necessary to allow it to resume synthesis. The polymerase backtracks, allowing an RNase to cleave the last few incorporated bases (which could potentially contain an error). Any necessary repair to the template may also occur at this time. The RNA polymerase then resumes its progression along the cleaned-up template. The alternative would be for the polymerase to abandon transcription of the gene until it was repaired. Some genes are extremely long, however, and significant effort would be wasted if this model were followed (for example, it takes nearly two hours to transcribe the 2-Mb *dystrophin* gene; having to start over after an hour and a half would be very wasteful of the cell's resources).

317. c.

The N-terminal region of free sigma protein folds over and shields the C-terminal region from binding to DNA promoter sequences; this binding is relieved when sigma factor complexes with core polymerase to form the RNA polymerase holoenzyme. The holoenzyme can thus recognize a promoter and initiate transcription properly.

318. a.

*Cis*-acting elements are DNA sequences that have an effect on the same DNA molecule on which they are located. A *trans*-acting factor is typically a protein produced by one gene that acts on a different DNA sequence.

319. c.

The *lacI* gene encodes the *lac* repressor protein, while the *lacA* and *lacY* genes encode transacetylase and lactose permease, respectively. All of these proteins are involved in the metabolism of lactose.

320. c.

Repair of cyclobutane pyrimidine dimers (CPD) is accomplished by several different mechanisms, depending on the organism in question. Many organisms have an enzyme called photolyase, which can cleave the cyclobutane linkage directly; other organisms such as humans lack this enzyme and must use a process called nucleotide excision repair to deal with these lesions. Phage T4 has the *denV* gene, which encodes an endonuclease that specifically cleaves the sugar-phosphate backbone of DNA at the site of a CPD.

321. b.

Plasmid DNA adopts three characteristic forms, which can be resolved by agarose gel electrophoresis. Form I is closed circular and supercoiled (both DNA strands are intact); Form II has a single strand break that has allowed the supercoiled stress to be relieved (the plasmid can be visualized as a hula hoop); Form III has a double stranded break and is thus a linear molecule.

322. d.

Constitutive genes, sometimes called housekeeping genes, have promoters that bind to general and upstream factors present in all cell types. They are thus expressed in all cells and at all times.

323. b.

In the helix-turn-helix, first identified in the λ repressor protein, one α-helix of the protein lies inside the major groove of DNA and can make site-specific contacts with the bases present there. A short turn of amino acids leads into a second α-helix that lies at an angle to the first; this second helix lies across the groove and makes contacts with the sugar phosphate backbone. This DNA binding motif is found in several DNA binding proteins in prokaryotes and eukaryotes.

324. d.

N and Q proteins are synthesized early in λ phage infection of *E. coli*. The presence of these proteins permits the transcribing RNA polymerase to read through a transcription "stop" signal and expressed downstream genes, leading to lytic infection.

325. False.

Children who are hemizygous have developmental abnormalities of the genitourinary tract as well as nephrogenic rests within both kidneys, according to this section of the listed reference: "The Knudson Model in Wilms Tumor."

326. a.

See "The WT1 Gene at Chromosome 11p13" in the listed reference.

327. e.

See Figure 24-6 in the reference.

328. e.

The WT1 gene protein does rise to a high at 20 weeks in the developing kidney but is barely detectable in the adult kidney. Also, the protein encoded by the *WT1* gene is expressed in the tissues described in *c* and *d*.

329. b.

See "Inactivation of WT1 in the Germ Line and in Wilms Tumors" in the listed reference.

The correct answers for questions 330–334 are given via definitions in the reference.

330. b.

331. a.

332. d.

333. c.

334. g.

The correct answers for questions 335–343 are given as definitions in the reference.

335. j.

336. b.

337. d.

338. c.

339. h.

340. a.

341. e.

342. f.

343. g.

344. b.

The first two genes expressed from the λ genome following infection are *N* and *cI*. The presence of N protein is required for RNA polymerase to read through the terminator at the end of the *cI* gene and into the *cro* gene.

345. a.

Cro protein binds first to the operator site $O_R3$, shutting down *cI* transcription from promoter $P_{RM}$, and then to site $O_R2$, shutting down *cI* transcription from promoter $P_R$. Repressor synthesis is thus inhibited and the infection proceeds to lysis.

346. c.

See Figure 29.4 in the reference.

347. b.

The greatly increased rate of sister chromatid exchange (SCE) in cultured BrdU-containing medium is the most striking aspect of Bloom syndrome.

348. e.

See "Disorders Associated with Trinucleotide Repeat Expansions" in the listed reference.

349. d.

See "Thrombosis and Factor V Leiden" in Chapter V of the listed reference.

The correct answers for questions 350–363 are given as definitions in the reference.

350. g.

351. b.

352. h.

353. l.

354. a.

355. c.

356. i.

357. d.

358. n.

359. k.

360. m.

361. f.

362. j.

363. e.

364. d.

See "Chromosome 17p: The p53 Gene" in the listed reference.

365. e.

See "Chromosome 17p: The p53 Gene" in the listed reference.

366. d.

See "Chromosome 18q: the DCC, SMAD4, and SMAD2 Genes" in the listed reference.

367. a.

Proteinase K is used to digest protein present in a sample.

368. c.

The purpose of RNase is to digest or destroy any RNA that may be present.

369. e.

Triton X-100 is a detergent.

370. e.

Changing the pH and the salt concentration elutes the DNA.

371. a.

The purpose of dialysis is to pull small molecules and salts from the area of higher concentration (sample on the membrane) to the area of lower concentration (buffer beneath the membrane).

372. d.

A ratio of 1.6–1.9 indicates that high-purity DNA has been isolated.

373. b.

TA ratio of < 1.6 indicates that the DNA is contaminated with protein, lipid, or other cellular components.

374. a.

Contamination with RNA is indicated by a ratio > 1.9.

375. d.

See product labeling in the New England Biolabs reference.

376. e.

See product labeling in the New England Biolabs reference.

377. a.

See product labeling in the New England Biolabs reference.

378. e.

See product labeling in the New England Biolabs reference.

379. True.

The addition of no enzyme results in no cutting and, thus, makes a good negative control to prove that your DNA sample is free of DNase enzymes.

380. d.

The 0.25 M HCl acts as a depurinating agent.

381. a.

The NaOH is added to denature the DNA in the gel.

382. True.

The presence of an air bubble or bubbles will not allow direct contact between the gel and the membrane, obviating the transfer of any DNA from the gel to the membrane at that point.

383. False.

Transferring the DNA from the gel to the membrane leaves no DNA in the gel to combine with the ethidium bromide and give fluorescence. Thus, at this point, the gel should <u>not</u> fluoresce if the transfer has been successful.

384. False.

Unlike nitrocellulose, which can autoignite in air at elevated temperatures, nylon is relatively stable and does not require a vacuum to keep the membrane away from air.

385. a.

The purpose of the enzyme terminal transferase (TdT) is to add a deoxynucleotides (in this case labeled with fluorescein) to the 3′ end of DNA. According to the product labeling, protruding, recessed, or blunt-ended double- or single-stranded DNA molecules serve as substrate for TdT.

386. d.

Areas on the membrane that may bind DNA non-specifically (i.e., bind DNA that is not target DNA) may be blocked with salmon sperm DNA.

387. False.

The statement is false since the presence of air bubbles prevents good contact between the hybridization buffer and the membrane.

388. d.

The fluorescein was added to provide a specific antigen rather then a detection probe by nature of its own fluorescence.

389. a.

The horseradish peroxidase is acting on its natural substrate to produce chemiluminescence, which is different from fluorescence.

390. e.

Lysine is still produced and the transition is, thus, "silent" even though the net effect is deleterious since splicing is affected.

391. c.

The correct answer is *c*, for the reason given in answer 390 above.

392. 98.8.

Although many formulae for estimation of the melting temperature exist, the one used is given below:

$$T_m = 81.5 + 16.6(\log M) + 0.41\ (\%G + C) - 0.61\ (\%form) - 500/L$$

where

$T_m$ is the melting temperature in °C.
M is the molarity of monovalent cations.
%G + C is the percentage of guanine and cytosine present in the DNA.
%form is the percent formamide in the solution.
L is the base pair length of the DNA.

393. d.

The 50S subunit of the prokaryotic ribosome has 31 proteins and 2 rRNA molecules: 23S and 5S. A long search for the proteins or combinations that possessed the peptidyl transferase activity has been unsuccessful. Extraction of 95% of the protein content leaves the 23S rRNA attached to protein fragments; this complex still has enzymatic activity. On the other hand, treatments that damage the 23S rRNA will abolish the transferase activity. In addition, 23S rRNA can catalyze the in vitro formation of a peptide bond between Ac-Phe-tRNA and Phe-tRNA, although the reaction is very slow. The role of the ribosomal proteins may thus be simply in precisely folding the RNA and/or positioning the reaction components such that the reaction can occur.

394. d.

During translation elongation, the Ef-Tu-tRNA complex binds to the A site of the ribosome. When it dissociates, EF-G binds and catalyzes the translocation of the ribosome to the next codon. Once the end of the mRNA is reached, a termination codon is now in the A site and so a tRNA does not bind; instead a release factor complex with similar shape and size binds, terminating translation as the growing peptide chain is instead transferred to $H_2O$. As all three of these complexes bind to the same A site, it is not surprising that they have similar shapes and sizes.

395. d.

See "PTEN Mutations on 10q23" in the listed reference.

396. a.

A solution of double-stranded DNA with a concentration of 50 μg/mL has an $A_{260}$ of 1.0. The diluted sample, therefore, has a concentration of 25 μg/mL. Because the solution was diluted 1:100 for measurement, the concentration is 2.5 μg/μL. The $A_{280}$ measurement is not necessary for this calculation.

397. b.

Clones E7 and F6 are separate, but each one overlaps with clone A1. Sequence information from clone A1 can therefore be used to connect the sequences from E7 and F6.

398. a.

The direction for reading the sequence of a DNA strand is always 5′-3′. By convention, the bottom strand is read from left to right and is also called the template strand, since it is the strand that codes for an mRNA. The top strand is read from right to left (also 5′-3′, due to the antiparallel nature of dsDNA) and is called the coding strand because it has the same sequence as the mRNA that is encoded (with the exception of U for T) in the mRNA molecule.

399. d.

The three main types of RNA are messenger RNA, ribosomal RNA, and transfer RNA. Complementary DNA is produced from reverse transcriptase and is abbreviated cDNA, but there is no such thing as cRNA.

400. b.

The term S refers to the Svedberg unit, a measure of how fast a particle will sediment in an ultracentrifuge. It has a rough relationship to mass, but is also highly dependent upon molecular shape. Hence the 30S and 50S components of the prokaryotic ribosome do not add up to 80S for the intact particle, but to 70S. The eukaryotic ribosome has subunit sizes of 40S and 60S, yielding an intact 80S particle.

401. b.

Because in the random primer reaction template DNA is denatured, random hexamers anneal to prime new DNA synthesis by Klenow fragment. Because DNA melting temperature depends upon the length of the double-stranded region, a hexamer sequence has a very low melting temperature. The hexamer will associate more stably with the template at room temperature than at 37 °C and remain annealed long enough for the DNA polymerase to extend. Klenow fragment is used because it lacks the 5′-3′ exonuclease activity that would cause one extending fragment to degrade a downstream fragment.

402. a.

Restriction endonucleases cleave the sugar phosphate backbone at a defined site. If DNA is synthesized using an $\alpha$-thio-dNTP, sulfate rather than phosphate is incorporated into the backbone; this alternative structure can be recognized by the enzyme, but cannot be cleaved.

403. c.

To prepare a riboprobe, one must use ribonucleotides. Many different types of labels can be used, from radioactive $^{32}P$ to non-isotopic biotin or digoxigenin.

404. a.

If only a single primer is added to an amplification reaction, the result is linear rather than exponential amplification. Taq polymerase will extend the primer and incorporate dU in place of dT opposite a template A, resulting in a radiolabeled, strand-specific probe.

405. a.

One of the characteristic features of DNA polymerase, whether it uses a DNA or an RNA template (reverse transcriptase), is that it is unable to synthesize DNA *de novo* as can RNA polymerase. Some form of primer with a 3′-OH group must be used. To primer RT-PCR, the initial primer can be gene specific, a random hexamer, or even an oligo-dT to primer mRNA amplification.

406. a.

Hybrid capture is a proprietary signal amplification method that uses antibodies directed against an RNA:DNA heteroduplex molecule. The epitope for the antibody is the physical structure of the heteroduplex. This antibody is used twice in the procedure: first it is bound to a solid matrix to capture target DNA, and secondly it is enzyme conjugated to allow the chemiluminescent detection of a captured heteroduplex.

407. g.

A thermostable DNA polymerase is required, as the procedure uses repeated melting and annealing steps; dNTPs are required because the primers that anneal are separated by one or more bases; extension of the upstream primer is required to make it adjacent to the downstream primer, after which ligase joins the two together.

408. c.

SSCP is not an amplification reaction, and ligase chain reaction requires thermal cycling.

409. a.

Nick translation uses the enzyme DNase I to generate random single strand breaks in dsDNA, while random primer labeling relies on primer annealing to ssDNA generated from the melting of dsDNA. Terminal deoxynucleotidyl transferase (TdT) uses as its preferred substrate ssDNA or dsDNA with a protruding 3′-OH; blunt-ended dsDNA or dsDNA with a recessed 3′-OH can be labeled by TdT if $Co^{2+}$ is included in the reaction as a cofactor. A riboprobe is the result of transcription from a promoter and is, therefore, the only method listed that will yield a single-stranded strand-specific probe.

410. b.

The SDA procedure is a non-isotopic method of amplification; it does not require a labeled nucleotide. However, it does need all of the other reagents listed.

411. d.

All of the choices are valuable information about the chromosomal fragments and thus allow the assembly of a completed chromosomal sequence following DNA sequencing.

412. a.

P-glycoprotein is a protein associated with excretion of digoxin, and the mutant form is less able to remove its substrate than is the wild-type protein.

413. Using the formula $(0.0114d - 0.154)^4$ where d = 5 cM, one should arrive at 0.0003%.

414. b.

Electrophoretic and vacuum transfer methods were developed as high efficiency alternatives to the simple capillary transfer method.

415. b.

DNA fragments longer than ~15 kb or so transfer very poorly. Soaking the gel in weak acid to cause partial depurination of the DNA, followed by treatment with strong base to neutralize and hydrolyze the phosphodiester backbone at the depurinated sites, reduces the fragment length and allows more efficient transfer.

416. d.

The correct answer is *d* because none of the other answers refer to *ADH1B*.

417. b.

Alkaline transfer to an uncharged nylon membrane does not fix the DNA in place; this method of transfer only accomplishes fixation if a positively charged nylon membrane is used.

418. b.

Increased concentrations of formamide help disrupt hydrogen bonding between base pairs and thus lower the $T_M$ of the duplex molecule. Increasing the concentrations of cations helps to shield the electrostatic repulsion between the negatively charged phosphodiester backbones and thus stabilize the duplex. Increasing the GC content of the probe (when feasible due to sequence constraints) or increasing the length of the probe serves to add more stabilizing hydrogen bonds, thus stabilizing the duplex.

419. c.

In a hybridization reaction, both the probe and the target molecules must be single stranded, otherwise annealing cannot occur.

420. a.

See the listed reference.

421. c.

In multiplex PCR, several PCR reactions can occur simultaneously using the same DNA template. By choosing primer pairs throughout the dystrophin gene, one can rapidly screen this large gene for exon deletions, one of the more common mutations found in Duchenne muscular dystrophy.

422. a.

A mutation that either creates or abolishes a restriction enzyme is called a restriction fragment length polymorphism. Restriction enzyme recognition sites are most frequently palindromic in nature—the enzyme binds the DNA as a dimer.

423. a.

Transfer RNA molecules contain several conserved features, among which are the length of the double-stranded regions forming the stems of the arms. The TψC arm of tRNA contains only four base pairs in its stem; it does however contain pseudouridine in the sequence TψC. Just 3′ to this arm is the acceptor stem that ends in the CCA to which the amino acid is attached.

424. c.

If a patient is heterozygous for a particular mutation, then the wild-type primer will amplify one allele and the mutant primer will amplify the other. Both reactions will therefore give a positive reaction.

425. b.

Trinucleotide repeat expansion is responsible for a number of neurological diseases, including fragile X syndrome, myotonic dystrophy, and Huntington disease. Cystic fibrosis is associated with more than 500 different polymorphisms in the CFTR gene, mostly involving codon deletions or substitutions.

426. b.

In reverse ASO hybridization the PCR product is labeled and used to probe a membrane spotted with oligonucleotides specific for the mutation of interest. This technique allows one to screen a single sample for several mutations simultaneously.

427. c.

In denaturing gradient gel electrophoresis, samples migrate through the gel through increasing concentrations of a denaturing agent. When they reach a critical concentration of denaturant they begin to melt; the sudden change in shape dramatically slows the migration of the nucleic acid through the gel.

428. a.

Decreasing the salt concentration will increase the electrostatic repulsion between the DNA strands, destabilizing the duplex and lowering the $T_m$. Conversely, increasing the salt concentration provides ample cations to shield the negatively charged phosphate groups of the backbone and stabilizes the duplex towards melting.

429. a.

NaCl will stabilize, rather than destabilize, the double helix. Formamide, heat, and urea disrupt hydrogen bonding, thus weakening the duplex.

430. i.

The production of a new strand of DNA from a DNA template, on the other hand, is called replication.

431. h.

Ethidium bromide is highly fluorescent when it intercalates between the base pairs of double-stranded DNA.

432. d.

Eukaryotic protein-encoding genes typically have their coding regions broken into translated regions called exons and untranslated regions called introns. Introns are spliced out during RNA processing to form the mature mRNA.

433. e.

The hybridization reaction can either take place in solution or on a surface such as a membrane.

434. f.

With its $pK_a$ of 7.2, MOPS is ideally suited to provide buffering power at a neutral pH. It is used frequently in RNA electrophoresis as it resists changes towards alkaline pH during the run, which could destroy the RNA.

435. g.

Other geometries of the double helix include the A-form (found in dsRNA and RNA:DNA heteroduplexes) and the left-handed Z-DNA.

436. c.

Uracil is not normally found in DNA but it is found in RNA. It arises spontaneously from the deamination of cytosine. If its presence is not detected, then replication will convert a GC base pair to an AT base pair. Uracil-N-glycosylase recognizes the base uracil attached to deoxyribose and removes it by cleaving the N-glycosidic bond between the base and the sugar. A second enzyme called AP endonuclease recognizes the sugar without an attached base and cleaves the phosphodiester backbone.

437. a.

In the hot-start PCR technique, a critical component of the reaction is kept out of the reaction until very high temperatures are reached and the wax bead is melted. This prevents aberrant primer annealing and extension as the temperature is increased and results in a cleaner product and a more efficient reaction. Many difficult-to-amplify sequences can only be amplified by this technique. Other methods for hot-start PCR include a specially modified *Taq* polymerase that is only activated following 10 min at 94 °C.

438. b.

Digoxigenin is a hapten that is conveniently incorporated into DNA during probe formation. It can be attached to a nucleotide base via a hydrocarbon linker arm that allows it to be attached but that keeps it from interfering with base pair formation. As a hapten it can be bound with an enzyme-conjugated specific antibody for its detection.

439. d.

Prokaryotic genes are found in clusters called operons; an operon is transcribed from a single promoter. The mRNA that is produced sometimes encodes several proteins; the first coding region always includes a Shine-Dalgarno sequence to properly orient the ribosome at the proper AUG start codon for translation. Downstream genes sometimes also include a Shine-Dalgarno sequence as well.

440. a.

The 5′ end of eukaryotic mRNAs has an inverted G residue at the end. This residue is not encoded by the gene, but is added post-transcriptionally during processing. Also during processing intron sequences are removed, exons are spliced together, and a poly-A tail is added to the 3′-end of the molecule.

441. b.

Ribosomes from prokaryotes are 70S in size, while those from eukaryotes are 80S. Each is comprised of two subunits, large and small, that contain rRNA and several proteins. The entire bipartite complex is the site of protein synthesis.

442. c.

The process of initiating eukaryotic translation is a complex one involving several different protein initiation factors. The ribosome assembles onto the mRNA in a sequential fashion, with the 40S subunit binding first to the 5′-cap site. This subunit then translocates down the mRNA until it encounters the start codon, an AUG in the proper surrounding context, after which the 60S subunit is brought into the complex. The ribosomal subunits are kept separate by eIF-3 until the final complex is ready to begin translation.

443. a.

Peptidyl transferase activity is thought to be primarily due to the rRNA found within the large subunit of the ribosome; ribosomal proteins are envisioned as having primarily a structural or positioning role, rather than a catalytic role.

444. d.

The release-factor complex, similar in shape to a charged tRNA, binds to the A site of the ribosome when a stop codon is reached. GTP cleavage is involved in the release of the finished polypeptide and the dissociation of the ribosome.

445. h.

Prokaryotic ribosomes contain 3 rRNAs—5S, 16S, and 23S—while eukaryotic ribosomes contain 4—5S (different from the prokaryotic version but similar in size), 5.8S, 18S, and 28S.

446. g.

DNA contains only AT and GC base pairs. RNA, with its flexible structure, can accommodate other non-Watson-Crick base pairs.

447. b.

During DNA replication, the DNA double helix separates into its strands. Each single strand serves a template for complementary DNA synthesis, resulting in daughter duplex molecules containing one parental strand and one newly synthesized strand.

448. c.

The early studies of Griffith and Avery, MacLeod, and McCarty (*Streptococcus*) and Hershey and Chase (phage T2) provided conclusive evidence that the transforming principle was DNA.

449. b.

The eukaryotic cell cycle is divided into four phases: M phase (mitosis) and S phase (DNA synthesis) that are separated by two gap phases called G1 (before S phase) and G2 (after S phase). Some cells also move from G1 into a resting state called G0; they can, under appropriate conditions, release from G0 and re-enter the cell cycle.

450. a.

Each strand of DNA is synthesized in the 5′-3′ direction as the phosphate group on the 5′ end of the incoming dNTP forms a phosphodiester linkage with the 3′-OH of the previous base.

451. c.

Eukaryotic chromosomes are linear, but the prokaryotic chromosome is usually circular. Replication proceeds bidirectionally from the single origin of replication in the bacterial chromosome.

452. b.

Slow stop mutants are those that can complete a round of replication that has already begun, but that cannot begin another.

453. a.

DNA polymerase III is the polymerase responsible for the replication of the genome, but DNA polymerase I (which exists in many more molecules per cell) has much higher activity. In fact, DNA pol III was not isolated until a mutant that had no Pol I activity was found.

454. d.

Nucleic acid synthesis always proceeds in the 5′-3′ direction, regardless of whether the enzyme is DNA polymerase, RNA polymerase, or reverse transcriptase.

455. a.

DNA polymerase III is a holoenzyme comprised of several subunits. The activities of all subunits are necessary for DNA replication, but the $\alpha$ subunit does the actual DNA synthesis.

456. b.

Whereas RNA polymerase can begin synthesis *de novo*, DNA must be synthesized from an already existing 3′-OH.

457. a.

One of the two daughter strands is synthesized continuously and one discontinuously.

458. d.

DnaB is the helicase that precedes the elongating complex, unwinding the DNA in front of the advancing polymerase; it is also involved in replication initiation by unwinding the duplex at the replication origin. DnaE is the $\alpha$ subunit that possesses the 5′-3′ polymerase activity, and DnaQ encodes the $\varepsilon$ subunit that is involved in the assembly of the holoenzyme dimer.

459. c.

During meiosis, homologous chromosomes line up and recombination can occur at this point.

460. d.

RuvA recognizes the structure of the Holliday junction and RuvB is an ATPase that functions as a hexamer; it has helicase activity that provides the motor for branch migration.

461. d.

Integration host factor (IHF) and Int are required for $\lambda$ phage integration, while IHF and Xis are required for excision from the *E. coli* genome.

462. d.

A retrovirus particle contains, typically, two copies of dsRNA, reverse transcriptase, and integrase.

463. d.

The typical retroviral genome contains at least three genes: *gag*, *pol*, and *env*. The *gag* gene encodes matrix, capsid, and nucleocapsid proteins; the *pol* gene encodes protease, reverse transcriptase, and Integrase proteins; and the *env* gene encodes surface and transmembrane proteins. Transducing viruses are replication-defective variants that have acquired cellular sequences; occasionally this includes an *onc* gene.

464. b.

As with all DNA polymerases, reverse transcriptase requires a primer with a free 3′-OH group to initiate DNA synthesis. In the case of retroviruses, the primer takes the form of an uncharged host cell tRNA molecule that is included in the virion as it is packaged.

465. b.

Retroviral proviruses initiate transcription from a single promoter located in the upstream long terminal repeat (LTR). Once transcribed, gag proteins are produced directly via ribosomal translation. Pol proteins are translated via the use of either a glutamyl-tRNA that suppresses the stop codon at the end of the *gag* gene or a ribosomal frameshift. *Env* genes are translated following a splicing event that generates a subgenomic RNA immediately following a portion of the gag message.

466. a.

Haploid *Saccharomyces cerevisiae* belong to either one of two different mating types, termed α and **a**. Some strains (those with a dominant *HO* gene) have the ability to change their mating type in response to certain signals. This mating type switching involves a specific recombination event.

467. a.

The mating type of a haploid yeast cell is determined by the genetic information present in the *MAT* locus, either α or **a**. Cells that carry the *MATa* allele are type **a**, while those that carry the *MAT*α allele are type α. In addition to the expressed information present at the *MAT* locus, silent copies of the α and **a** sequences are present in flanking loci called *HML* (left of the *MAT* locus, containing α sequence) and *HMR* (right of the *MAT* locus, containing **a** sequence).

468. a.

The gene *HO* encodes an endonuclease that catalyzes the mating type switching recombination event in haploid yeast. The gene is not expressed in diploids because the cell is expressing both α and **a** genes already.

469. d.

Eukaryotic DNA is organized into linear chromosomes; each one contains a centromere, telomere sequences at the ends, and at least one origin of replication (although frequently there are many).

470. b.

The haploid genome size is three billion base pairs. Assuming that the DNA exists totally in the canonical B-form with a helical pitch of 3.4 Å per base pair, the diploid genome found in a typical human cell would stretch just over two meters in length if laid end-to-end.

471. d.

The nucleoid is the term for the region in a bacterium that contains the circular chromosome; it is a region of the cytoplasm, because prokaryotes do not have a true nucleus.

472. a.

Heterochromatin is typically found at the centromeres of chromosomes, but can be found in other regions as well. It is very tightly packaged with proteins, and this packaging changes little throughout the cell cycle.

473. h.

The core histone octamer consists of a tetramer of $H3_2 \cdot H4_2$, flanked on top and bottom by two dimers of H2A•H2B. Approximately two turns of DNA are wrapped around this protein core; the entry and exit points are bound by a single molecule of H1 to form the nucleosome.

474. a.

Histones are highly basic proteins, having a large number of lysine and arginine residues. They therefore bind strongly to negatively charged DNA.

475. b.

A nucleosome contains approximately 200 bp of DNA (146 bp if just the core particle is considered). Each base pair has two negatively charged phosphate groups, and so is highly negative. Positively charged histones shield some of these charges, but the overall charge of the nucleosome is still highly negative.

476. e.

The path of the DNA around the core octamer is approximately circular, but subtle variations in this path can be observed with high resolution mapping (for example, hydroxyl radical footprinting) techniques. Answers *b* and *c* are also not true.

477. d.

DNA that is packaged into nucleosome structures is relatively protected from exposure to enzymatic digestion due to steric hindrance. When a gene is activated, regions of the gene often have their nucleosomes either removed or disrupted such that this protection is lost; the region thus becomes accessible to nucleases such as DNase I.

478. a.

Activation of a gene requires that the promoter sequence be made accessible to transcription factors. The canonical nucleosome structure is therefore removed; histone proteins may still remain bound in the region but the overall nucleosome structure is altered such that the DNA is dramatically more accessible to other protein factors.

479. h.

TATA-binding protein is a component of transcription factors for all three types of RNA polymerase. Its mode of DNA binding is relatively unusual, in that it binds in the minor groove instead of the major. When it binds to DNA it introduces an ~80° bend, which helps other transcription factors and RNA polymerase to bind.

480. c.

TFIIIC is an assembly factor for the initiation of transcription of the 5S gene by RNA Pol III; when it binds it directs TFIIIB to bind at the proper location. TFIIIA is another assembly factor for Pol III transcription.

481. d.

The phosphorylated CTD of RNA Pol II also binds the enzyme that is responsible for modifying the 5′ end of the newly synthesized mRNA. During transcription initiation, RNA polymerase II binds to the promoter in a non-phosphorylated state followed later by factor TFIIH. TFIIH has several activities, one of which is a kinase activity that phosphorylates the CTD of RNA polymerase II, prompting the release of several factors and allowing the transcription process to begin.

482. c.

Bacterial sigma factor binds to the AT-rich –10 sequence and is required to initiate transcription at promoters. In eukaryotes, the TATA binding protein (TBP) binds to this sequence, termed the TATA-box, which is located about 35 bp upstream of the start point.

483. c.

RNA Pol I transcribes the 5.8S, 18S, and 28S rRNA genes, and RNA Pol II transcribes all protein-encoding genes (mRNA).

484. b.

The correct answer is *b* because a site with dyad symmetry is one in which the base sequence when read in the 5′-3′ direction is identical on both strands; these sites typically are bound by protein dimers.

485. b.

If intron sequences contain significant regions of complementarity, they can interact and the splicing reaction can join two separate molecules, rather than the canonical *cis*-joining. Although a rare event, it occurs *in vivo* in some special situations, such as in numerous trypanosomal mRNAs, in the *C. elegans* actin genes, and in chloroplasts.

486. b.

While the mouse λ locus contains only two V and four J-C segments, the human homologues contain ~300 V and more than six J-C segments. D segments are not found in the light chain genes.

487. b.

Klenow DNA polymerase is the large fragment of *E. coli* DNA polymerase I that is produced by proteolytic digestion with the enzyme subtilisin. It contains the 5′-3′ polymerase and 3′-5′ proofreading exonuclease activities, but the 5′-3′ exonuclease activity has been lost.

488. c.

The 3′-5′ exonuclease activity is termed "proofreading" because it will remove the base just inserted by the polymerase activity in case of a mismatch. The enzyme will then resume synthesis, this time including the correct nucleotides. This gives replication a very high fidelity.

489. a.

Many reverse transcriptase enzymes also contain an associated RNase H activity, to degrade the RNA strand in the newly synthesized RNA:DNA heteroduplex molecule.

490. c.

DNase I is a small enzyme and as such it has relatively easy access to DNA. The presence of bound proteins or even alterations of DNA structure are sufficient to alter the frequency of DNA cutting, and so it is frequently used to map positions of protein binding.

491. c.

DNA ligase requires as substrates a 3′-OH, a 5′-phosphate, and ATP.

492. c.

Each of these enzymes is used to produce a labeled DNA fragment. Klenow fragment is used in random primer labeling and for filling in a 5′-overhang; TdT adds nucleotides to the 3′ end of ssDNA or dsDNA with a 3′ overhang; T4PNK transfers the terminal phosphate group of ATP to the 5′-OH of DNA; alkaline phosphatase removes a 5′-phosphate group in preparation for end-labeling with PNK.

493. a.

Bacterial artificial chromosome (BAC), based on the naturally occurring F plasmid, can accept insert sizes of 300 kb or longer. Some yeast artificial chromosomes can accept inserts of 600–1400 bp, but these are less useful because they suffer from insert rearrangements.

494. d.

X-gal is a colorless compound that can be cleaved by the enzyme β-galactosidase; when cleaved it produces a blue precipitate. Preparing recombinant plasmid involves cloning an insert into, and thus disrupting, the β-galactosidase gene. The lack of a functional β-gal gene leads to the production of white colonies on the plate; blue colonies indicate that the insert was not ligated into the plasmid.

495. d.

The template for any DNA polymerase is ssDNA with an annealed primer providing a 3′-OH group; the ssDNA ahead of the primer provides the template for new DNA synthesis.

496. a.

In the PCR process, template DNA is first heated to denature the double helix. Second, the temperature is lowered to allow annealing of gene-specific primers. The temperature is then increased to allow Taq polymerase to extend the annealed primer to the end of the template. Repeated cycling of these three steps produces exponential amplification.

497. b.

This is the definition of a restriction fragment length polymorphism. Creation or abolition of a recognition site will lead to the production of differently sized fragments following electrophoresis. Analyzing the patterns produced facilitates the construction of a rough genomic map.

498. c.

There are four possible bases at any given position in a DNA sequence. In random sequence DNA the probability of a particular six-base combination is thus $(1/4)^6$, or 1 in 4096.

499. a.

Simple sequence length polymorphisms (SSLs) are arrays of repeat sequences that display length variations, different alleles containing different numbers of repeat units. There are two types of SSLP: minisatellites, also known as VNTRs, and microsatellites, also known as STRs.

500. b.

Somatic cell hybridization was among the first tools used to study the genetics of mammalian cell differentiation. Infection by the Sendai virus causes somatic cells to fuse, producing a heterokaryon, a nucleus containing chromosomes from the two different cell lines. Cellular fusion can also be induced chemically by treatment with polyethylene glycol. Fused cells are then screened for uptake by plating them on selective media such as HAT; the hamster cell line chosen for fusion is deficient in the enzymes thymidine kinase and hypoxanthine phosphoribosyl transferase and thus will die unless they have taken up DNA fragments containing these two genes.

501. e.

Northern blots, RT-PCR, and RNase protection assays are all commonly used methods for studying gene expression using RNA isolated from cells or tissue. Southern blot analysis, on the other hand, is a technique that examines genomic DNA and not RNA.

502. d.

*In situ* hybridization is a technique that usually involves the hybridization of a single-stranded anti-sense RNA probe to a specific mRNA found in the cytoplasm of cells of interest.

503. d.

In situ hybridization enables the detection of the specific cells that have a particular mRNA present in the cytoplasm. One of the major benefits of this technique is that it preserves the histology of the tissue while still enabling hybridization of a specific probe to the mRNA of interest.

504. True

See the answer above for further information.

505. d.

Riboprobes are defined as single-stranded RNA probes. These probes are usually anti-sense probes, meaning they are complementary to a specific mRNA and will hybridize to that mRNA.

506. c.

RNA interference (RNAi) is a method for degrading a specific mRNA transcript in the cytoplasm before it can be translated.

507. c.

RNAi depends on the activity of dsRNA in the process; ssRNA can serve to post-transcriptionally block translation, but it is much less effective.

508. c.

Dicer is the name given to the enzyme found in cells that produces the short dsRNA fragments that go on to form the RISC complex.

509. b.

The name of the protein complex that forms with the short 21–25 base pair dsRNAs is referred to as the RISC (RNA-induced silencing complex).

510. a.

RISC functions by cleaving the target mRNA, thereby preventing it from being translated.

511. c.

The term siRNA refers to short interfering RNA (also called small interfering RNA) that are complementary to the target mRNA. When introduced into cells, these short dsRNA molecules (generally 20–25 base pairs) can also be effective in inducing RNA interference.

512. b.

Work with mammalian cells indicates that the siRNAs are more effective than the longer dsRNA molecules typically used in RNAi in *C. elegans*, for example.

513. a.

Gene knockouts are expensive and very time consuming to perform. Another potential advantage to RNAi is that the effect is temporary, while the effect of a gene knockout is permanent.

514. e.

RNase protection depends upon a single-stranded RNA probe hybridizing to the target mRNA. Thus, the ssRNA probe must be complementary to the target mRNA.

515. d.

EcoRI and DNaseI act on DNA, not RNA. Ribonuclease P does not have the appropriate enzymatic activity required (ability to degrade ssRNA but not dsRNA). Of the choices given, only Ribonuclease A has the appropriate enzymatic activity.

516. a.

In RNase protection assays, the dsRNA remaining after the RNase digestion is run on a gel, and the amount of RNA in the band, are quantified relative to standards run on the same gel.

517. d.

Answers *a*, *b*, and *c* are all examples of uses of RNase protection experiments in studying gene expression.

518. d.

In DNA fingerprinting analysis, the repeated sequences in the human genome that are amplified by PCR are short tandem repeats (STRs).

519. a.

Multiplex PCR is the term given to PCR that is conducted using multiple pairs of primers to amplify multiple targets in the same reaction tube.

520. b.

Individuals vary at the STR sites by the number of copies of the short tandem repeat at that location. Thus, the differences between individuals are seen as sequence-length polymorphisms.

521. c.

DNA fingerprinting using automated DNA sequencing instruments uses fluorescent-labeled primers. This is because the automated instruments are designed to read fluorescent labels using specialized lasers and detectors.

522. b.

No two individuals are alike at all 13 sites. However, any two individuals are likely to be the same at one or more of the sites.

523. c.

The technique 5′ RACE (rapid amplification of cDNA ends) was developed to help identify the start point for transcription. This is useful because it helps researchers find the promoter sequence that should be located immediately upstream of the genomic sequence that encodes the 5′ end of the transcript.

524. a.

Nuclear runoff assays are designed to measure how much of a particular mRNA transcript was actively being transcribed in cells at a specific time. Other methods of measuring a particular mRNA from cells, such as Northern blots and ribonuclease protection assays, measure the amount of accumulated mRNA, not the amount being actively transcribed.

525. d.

Affymetrix arrays use 20–25 long oligonucleotides, while other types of arrays use much longer oligonucleotides (50–75 long) or cDNAs.

526. d.

Reverse transcriptase is the only enzyme that can synthesize DNA from an RNA template.

527. b.

Even though the other types of labels are theoretically possible, virtually all microarrays use fluorescent probes.

528. c.

The device used to "read" the fluorescence on a microarray images the fluorescence using confocal laser light and appropriate detectors.

529. b.

The A probe is from the normal cells and the B probe is from the cancer cells. Therefore, a spot that hybridizes with the A probe but not the B probe represents a gene that is expressed in normal cells but not cancer cells.

530. c.

The A probe is from the normal cells and the B probe is from the cancer cells. Therefore, a spot that hybridizes with the B probe but not the A probe represents a gene that is expressed in cancer cells but not normal cells.

531. a.

The A probe is from the normal cells and the B probe is from the cancer cells. Therefore, a spot that hybridizes with the A probe and the B probe represents a gene that is expressed in both normal cells and cancer cells.

532. d.

A spot that hybridizes with neither the A probe nor the B probe represents a gene that is not expressed in normal cells or cancer cells.

533. d.

Photolithography is the patented process developed by Affymetrix to synthesize thousands of oligonucleotides simultaneously on a microarray chip.

534. d.

The four possible alternative spliced transcripts are: (1) exons 1, 2, and 3; (2) exons 1 and 2; (3) exons 2 and 3; and (4) exons 1 and 3. No rearrangements in splicing are possible; that is, an arrangement such as exon 3 followed by exon 1 is not possible.

535. b.

The splicesome is the macromolecular structure that consists of both proteins and RNA that splices pre-mRNA to mRNA.

536. e.

Answers *a*, *b*, *c*, and *d* are all true statements.

537. c.

Primers provide the specificity of a PCR reaction. If the primers are properly designed and if an appropriate annealing temperature is chosen, then the primers should hybridize only to the target sequence that is complementary to the primers.

538. a.

The great sensitivity of PCR makes false positive results a serious concern. Great care must be taken to prevent contamination of PCR reactions with exogenous DNA from the environment.

539. b.

When amplifying genomic DNA from

- very old samples,
- samples that are likely to be degraded due to improper storage,
- samples preserved using methods not optimal for DNA storage, or
- samples collected from the environment such as forensic samples,

amplification of very short target sequences greatly increases the probability of a successful PCR. This is because the shorter the amplification target, the more likely that both primer sites will be present on at least some of the degraded DNA fragments.

540. e.

Despite the movie "Jurassic Park," dinosaur DNA has not been successfully isolated from ancient insects found preserved in amber.

541. e.

Answers *a*, *b*, *c*, and *d* are all true and their use in crime scene investigations is well documented.

542. a.

This kind of work using preserved cancer tissue has been an important research tool for many years. The method can also be applied to tissue that has been stored for other purposes.

543. c.

"Long PCR" protocols use two different thermostable DNA polymerases. Typically, one of these enzymes has the $3'-5'$ exonuclease function (proofreading) while the other enzyme does not.

544. c.

545. a.

546. d.

An EST is derived from a single sequencing reaction performed on a cloned cDNA molecule. They are called "expressed sequence tags" because, as a cDNA molecule, the sequence is derived from mRNA and therefore was expressed in the cells used in the mRNA isolation to construct the library. They are "sequence tags" because the sequence is a partial one, but provides enough sequence information to serve as an information "tag" for that gene.

547. b.

The non-redundant data set (nr) includes all genomic and cDNA sequences collected from all organisms. EST data is found in a separate database called dbEST.

548. a.

An open reading frame is a region of genomic DNA that has no stop codons in one of the three reading frames. The minimum length of the region of genomic DNA that is considered an "open reading frame" can be set by the investigator.

549. b.

With the advent of complete genome sequencing projects, Genefinder programs have become extremely important in the analysis of genomic DNA. Imagine you have the complete sequence of the human genome sitting in front of you (all three billion base pairs of A, G, T, and C). How would you identify where the genes are in that sea of DNA? Keep in mind that the coding region of our genome makes up about 1% of the total DNA. Also remember that our genes typically consist of many small exons separated by very long introns. The problem of finding genes is a non-trivial one and no two Genefinder programs will agree when examining a complete genome.

550. d.

Although the other answers are possible, gene fusions most often involve the fusion of one gene's promoter to a "reporter" gene (see the next question for more detail).

551. e.

Green fluorescent protein (GFP), firefly luciferase, and beta-galactosidase are all examples of commonly used reporter genes.

552. e.

In order to get high levels of protein production in *E. coli*, the expression vector must contain all necessary signals for both transcription and translation in *E. coli*. These signals include a strong promoter for transcription and a ribosome binding site for translation. In addition, restriction sites for cloning the gene of interest must be present directly downstream from the promoter and ribosome binding site.

553. b.

Nickel bound to a column is bound by the "six His tag" at the beginning of the fusion protein.

554. e.

Answers *a*, *b*, and *c* are all correct.

555. a.

Pulsed field gel electrophoresis (PFGE) is an electrophoresis technique that can separate DNA fragments up to 9 Mb (9000 Kb) in length!

556. c.

Electroporation is the most efficient technique currently available for getting such large molecules into *E. coli* cells.

557. b.

The term for selecting those cells that have taken up the plasmid from those that have not is "selection for transformants."

558. c.

Random shearing by nebulization is the current method of choice to get fragments for cloning in the right size range. The 8-cutter enzymes cleave the DNA into very large fragments (an average >60 Kb), while 4-cutter enzymes cleave the DNA into very small fragments (about 250 base pairs). DNaseI cleaves DNA randomly but is too difficult to control to get the right fragment sizes.

559. d.

Three, six, or twelve billion base pairs are not nearly enough. Typically, a genome project must sequence seven to ten times the number of base pairs in the genome in order to assemble the entire sequence. Even then, small gaps will exist that will need to be selected and sequenced using a variety of methods.

560. b.

At that time there were still many gaps in the human genome sequence. They were celebrating the announcement of a completed first draft of the sequence that likely included about 90% of the genome.

561. d.

Answers *a*, *b*, and *c* are all correct. All of these different kinds of gene information are extremely important in identifying the location of genes in a newly sequenced genome.

562. a.

In most gene annotations, this is the critical information that is provided. Often the precise location of the promoter and other regulatory elements are not known. The 3-D structure of the protein is also often not known.

563. g.

Answers *a*, *b*, *c*, *d*, *e*, and *f* are all correct. These are all important components of yeast artificial chromosome vectors.

564. a.

Cosmids are designed with cos sites from bacteriophage lambda so they can be packaged and introduced very efficiently into *E. coli*. Electroporation is not nearly as efficient, and lipofection and calcium phosphate precipitation are not used for *E. coli*.

565. c.

*Eco*RI methylase protects the *Eco*RI restriction sites by methylation, thereby preventing cutting by *Eco*RI restriction endonuclease.

566. Bacterial cells possess restriction/methylation systems because they protect their own DNA with the methylase enzymes, but use the restriction enzymes to cut the DNA of invading bacteriophage. Thus, the bacterial restriction/modification systems can be viewed as a method for distinguishing "self" from "non-self."

567. b.

Recent research has shown that histone modification is important in gene regulation in eukaryotes.

568. b.

Apromoter is defined as a cluster of functionally related genes that are transcribed together on a single mRNA from one promoter. The separate coding regions on the mRNA are translated separately into different proteins (i.e., not a polyprotein).

569. e.

Both *b* and *c* are correct (see below).

570. e.

The *lac* operon has three genes under the control of a single promoter: *lac*Z, *lac*Y, and *lac*A.

571. a.

The binding site for the *lac* repressor is the *lac* operator sequence. When the *lac* repressor is bound to this sequence, the RNA polymerase is incapable of moving off the promoter and transcribing the genes of the operon.

572. a.

In the absence of lactose there is no need to transcribe the genes of the *lac* operon and, thus, *lac* repressor protein is bound to the *lac* operator. Upon binding of lactose by the bound *lac* repressor, conformational change occurs in the *lac* repressor protein, rendering it incapable of binding to the *lac* operator. Transcription of the *lac* operon then occurs, yielding the proteins necessary for lactose metabolism. Thus, the presence or absence of lactose is the key to this regulation.

573. f.

The regulation is negative because when the *lac* repressor binds the operator, the effect is to shut down transcription. The regulation is inducible because the presence of lactose serves to "induce" expression of the operon.

574. c.

The key to glucose regulation is the fact that as glucose levels decrease in the cell as the glucose in the environment is used up, cAMP levels in the cell increase. As the cAMP levels increase, cAMP binds to the CAP protein, which can then bind to the *lac* promoter. The *lac* promoter by itself is a weak promoter, but with CAP—cAMP bound, the *lac* promoter now becomes a strong promoter and binds RNA polymerase very efficiently.

575. a.

The enzyme adenyl cyclase is responsible for converting ATP to cAMP in *E. coli*.

576. b.

*E. coli* uses the glucose more efficiently than lactose, so it makes sense for glucose to be involved in repressing the *lac* operon.

577. a.

When glucose is absent, cAMP is produced and binds to CAP protein, CAP protein binds to the *lac* promoter, and increases transcription of the *lac* operon. When a regulatory protein binds to the DNA and increases transcription, it is defined as positive regulation.

578. d.

It is repressible, because biosynthetic operons are repressed by the presence of the small effector molecule (in this case tryptophan).

579. b.

Tryptophan is called the co-repressor because it binds to the repressor protein to repress synthesis of the *trp* operon.

580. c.

The *trp* repressor protein (when it is bound to W) binds to the *trp* operator sequence to repress transcription.

581. d.

The form of regulation described in the question is referred to as attenuation because, when lots of tryptophan is present in the cells, transcription of the operon is attenuated before the genes of the operon are transcribed.

582. b.

Attenuation of the *trp* operon occurs when levels of tryptophan in the cell are high.

583. b.

*Salmonella typhimurium* is the species used in the original version of the Ames test.

584. e.

Both his- missense and his- frameshift strains are used to screen for mutagenic agents in the Ames test.

585. d.

The data clearly show that cigarette smoke condensate is causing significant numbers of frameshift mutations but not missense mutations. Keep in mind that this question is hypothetical and that in real experiments, cigarette condensate has been shown to contain both missense and frameshift mutagenic substances.

586. d.

Transformation, conjugation, and transduction are all different routes for the introduction of DNA into bacterial cells. Transcription is the process by which DNA is copied into RNA, while translation is the process by which the information in mRNA is converted into an amino acid sequence of a protein.

587. e.

*E. coli* cells must be treated in the laboratory with special solutions in order to become *competent* for transformation.

588. c.

The process of bacterial conjugation is often referred to as bacterial mating.

589. b.

The F plasmid (or fertility factor) encodes the genes necessary for *E. coli* conjugation.

590. a.

The F plasmid (fertility factor) has genes that encode the sex pilus. The sex pilus is the tube used to transfer DNA from one cell to another during bacterial conjugation.

591. e.

Answers *a*, *b*, *c*, and *d* are all true about plasmids.

592. b.

R factors are defined as plasmids that are referred to as resistance factors that carry one or more antibiotic resistance genes.

593. c.

After several dysentery epidemics in Japan in the late 1950s, the pathogen was found to be resistant to several antibiotics due to the antibiotic resistance genes carried on plasmids.

594. a.

Horizontal gene transfer is defined as the transfer of a gene from one species to another species.

595. d.

Transposons are short segments of DNA that can move or be "transposed" from one location to another.

596. c.

Transposons occur in both eukaryotes and prokaryotes.

597. c.

Barbara McClintock discovered transposons in corn in the 1950s.

598. d.

The simplest transposons in prokaryotes are referred to as insertion sequences (IS).

599. d.

Both *a* and *b* are correct.

600. d.

Answers *a*, *b*, and *c* are all correct.

601. a.

Transduction is the process whereby bacterial DNA is transferred from one cell to another via a bacteriophage.

602. e.

Avery, MacLeod, and McCarty demonstrated that DNA was the genetic material by transforming avirulent, nonencapsulated bacteria to virulent, encapsulated bacteria.

603. d.

An Hfr (high frequency recombinant) cell is one that has the F factor integrated into the bacterial chromosome. During mating, such an Hfr cell can transfer part of the bacterial chromosome to a recipient F- cell.

604. e.

An F- cell is one that has no F-factor (F plasmid).

605. b.

An F′ cell is one that has an F plasmid carrying a piece of DNA from the bacterial chromosome.

606. a.

When an F+ cell mates with an F- cell, a copy of the F plasmid is passed to the F- cell, converting it to an F+.

607. c.

The spread of plasmids via bacterial conjugation is the most important route of antibiotic resistance gene transfer between bacterial strains and species.

608. e.

All of the cell types listed have been used to produce recombinant drugs to treat human disease.

609. h.

The items in answers *a*, *b*, *c*, *d*, and *e* are all products produced in *E. coli* for use in treating human disease.

610. h.

Items *a*, *b*, *c*, *d*, and *e* are all products produced in *Saccharomyces cerevisiae* (Baker's yeast) for treating human disease.

611. e.

Of the products listed, only superoxide dismutase has been produced in *Pichia pastoris*.

612. h.

All of the products listed have been produced in mammalian cells in culture.

613. c.

Antitrypsin is produced in genetically modified sheep.

614. f.

Adenoviruses, retroviruses, and plasmids have all been used in gene therapy efforts.

615. b.

The *p53* gene encodes a tumor suppressor protein.

616. b.

The estimated number of genes in the human genome is 30,000 to 40,000 genes.

617. c.

The polymerase chain reaction was crucial to amplifying the 16S rRNA genes of the bacterium in order to make an identification. This method is often used to amplify the DNA of organisms that can't be cultured in the laboratory.

618. d.

A Hanta virus outbreak in 1993 in the American Southwest was identified rapidly by PCR.

619. a.

The new tick-borne disease identified in 1994 was caused by *Ehrlichia chaffeensis.*

620. a.

The first PCR-based diagnostic test approved by the Food and Drug Administration was for *Chlamydia trachomatis*.

621. ssRNA

622. ssRNA

623. dsRNA

624. ssRNA

625. dsDNA

626. ssDNA

627. ssRNA

628. dsDNA

629. ssRNA

630. ssRNA

631. ssRNA

632. dsDNA

633. ssRNA

634. ssRNA

635. ssRNA

636. dsDNA

637. ssRNA

638. ssRNA

639. ssRNA

640. ssRNA

641. ssRNA

642. ssRNA

643. c.

An RT-PCR test was developed in 1999 to amplify and identify the RNA genome of the West Nile encephalitis virus.

644. e.

All of the viruses listed are oncogenic viruses with DNA genomes except for HTLV-1, which has an RNA genome.

645. d.

All of the viruses listed are oncogenic viruses with RNA genomes.

646. e.

The causative agent of Creutzfeldt-Jakob disease is thought to be a prion, an infectious agent that is pure protein with no nucleic acid genome.

647. a.

Scrapie is a prion disease of sheep, not humans.

648. b.

Viroids are short pieces of naked RNA 300–400 nucleotides long that cause some important plant diseases (such as potato spindle tuber viroid).

649. f.

These giant bacteria have no membrane-enclosed nucleus and have flagella that resemble the bacterial type and not the eukaryotic type. These cells are larger than many eukaryotic cells and do not reproduce by binary fission. Instead, daughter cells are formed within the parent cell and are released through an opening in the parent cell. In 1999, an even larger species of bacterium was discovered in sediments off the coast of Namibia (*Thiomargarita namibiensis*).

650. e.

Only *a* and *b* are true: Archaea are not sensitive to the same antibiotics as the true Bacteria.

651. b.

The membrane lipids of Archaea are indeed composed of branched carbon chains attached to glycerol by ether linkages. However, Eukarya membrane lipids are composed of straight carbon chains attached to glycerol by ester linkages. The common tRNA described in answer *c* is common to Bacteria and Eukarya but not Archaea.

652. e.

Pathogenicity islands are large blocks of DNA that carry virulence genes. These genes appear to be transferred between species via horizontal gene transfer. Answer *c* is incorrect because a pathogen may have more than one pathogenicity island.

653. b.

Type III secretion in gram-negative bacteria involves the injection of proteins into the host cell. These proteins resemble the eukaryotic factors involved in cell signal transduction. They interfere with normal eukaryotic signaling pathways and may interfere with host immune responses.

654. f.

Pathogenic bacteria of all of the listed species carry pathogenicity islands.

655. a.

Exotoxins are almost always encoded by genes on plasmids or prophages, while endotoxins are encoded by chromosomal genes.

656. b.

Exotoxins can be secreted by both gram-positive and gram-negative bacteria and are highly toxic (botulinum toxin is a good example). Endotoxins are found as part of the outer membrane of some gram-negative bacteria and are generally of moderate toxicity.

657. c.

658. g.

659. h.

660. e.

661. g.

662. h.

663. i.

664. a.

665. f.

666. j.

667. a.

668. d.

669. i.

670. b.

671. This statement is true. Many antibiotic resistance genes encode proteins that directly break down or modify the antibiotic so that it no longer functions.

672. d.

673. a.

674. b.

675. b.

676. e.

677. f.

678. c.

679. b.

680. c.

In ribotyping, the chromosomal DNA from many strains is typically cleaved with a restriction endonuclease. The DNA is then run on a gel and transferred in a Southern blotting procedure to a membrane. The membrane is then hybridized with a nucleic acid probe that will bind to the ribosomal genes on the membrane. The resulting patterns of bands are different for different strains.

681. b.

In plasmid fingerprinting, plasmids are isolated from a microbial isolate and the banding pattern is examined by agarose gel electrophoresis.

682. c.

Selenocysteine is attached to a tRNA that recognizes the UGA codon with the help of specific translation factors.

683. True

The UGA codon is not a stop codon in mitochondria and mycoplasmas.

684. True

Some archaea (and possibly some bacteria) can incorporate pyrrolysine opposite UAG (which is usually a stop codon).

685. False

Archaea use the same genetic code as bacteria and eukaryotes. However, AUA does specify methionine in human mitochondria.

686. True

UAG binds a tRNA carrying glutamine in some protozoans and algae. Remember, UAG also specifies pyrrolysine in some prokaryotes.

687. False

CGG encodes tryptophan only in plant mitochondria.

688. False

N-formylmethionine is encoded by AUG only in bacteria (and in mitochondria and chloroplasts) but not in eukaryotes and archaea.

689. c.

Beta-lactamase cleaves the penicillin and renders the bacterial cell resistant to the antibiotic.

690. d.

The activated receptor breaks down the repressor of the beta-lactamase gene. With the repressor degraded, the beta-lactase gene is transcribed and the protein is produced.

691. d.

All of the hypothetical drugs could theoretically prevent the production of beta-lactamase, rendering the cells sensitive to penicillin.

692. d.

Items *a*, *b*, and *c* would all be necessary to use this as a reasonable approach to help combat human diseases like malaria. Malaria, for example, kills more than two million people worldwide each year, so the motivation to successfully achieve this type of work is high. However, environmental/ecological issues are a great concern due to worries about releasing genetically modified insects into the environment.

693. a.

Trials in laboratory cattle have been successful and a field trial is currently ongoing.

694. c.

Ciprofloxacin (or "Cipro") was distributed by the U.S. government (the government bought 100 million Cipro tablets).

695. c.

Adenovirus has been used in trials to investigate the use of viruses to introduce normal genes into the cells of patients with a genetic disease.

696. d.

Each HIV virus has two single-stranded RNA genomes (+ strand RNA) and two tRNA molecules.

697. e.

All of the statements regarding the copying of the ssRNA of a retrovirus to dsDNA are true.

698. c.

Simian immunodeficiency virus has a nucleotide sequence that is closest to that of HIV.

699. b.

Integrase is the enzyme carried within the virus that is responsible for integrating the ds HIV DNA into a host chromosome.

700. a.

The tRNA crucial to HIV replication is encoded and synthesized in the human cell in which the virion was packaged.

701. True.

The amino terminus of the protein is derived from the 5′ end, which is the promoter-enhancer region, while the carboxy terminus is derived from the 3′ end where the poly-A "tail" of the mRNA resides.

702. e.

The substitution of a G for a C at nucleotide position 3586 results in the replacement of aspartic acid (D) by histidine (H) at amino acid 1152 (D1152H).

703. False.

Standard ASPCR gives false positives for D1152H in the presence of the wild-type maternal allele.

704. a.

The assay was designed to detect the D1152H mutant from the paternal gene in the presence of a large amount of maternal DNA that is wild-type.

705. False.

Fetal DNA represents only about 3.4–6.2% (Lo et al. in reference 32 of the paper) of the total cell-free DNA circulating in maternal blood.

706. d.

Pre-eclampsia is one disease that will raise the plasma total DNA. Also, certain viral infections can raise the blood cell count, which can raise the total plasma DNA through cell leakage and outright lysis. Likewise, delaying the processing of whole blood or mishandling can lead to a higher plasma DNA due to the lysis of DNA-containing cells in whole blood.

707. c.

This is the stated level of maternal wild-type DNA up to which a true positive can be obtained in the high-accuracy assay, according to the authors.

708. e.

It is the stated purpose of the "housekeeping gene" assay to determine the total DNA in the maternal samples.

709. a.

This is the difference emphasized by the authors.

710. b.

According to the reference, the main purpose of the TaqMaster PCR Enhancer is to reduce possible inconsistencies in PCR amplification. The secondary purpose is this: the authors found that including the additive helped decrease the formation of primer-dimer artifacts, which can compete with the desired PCR product and reduce its yield.

711. c.

This is the stated purpose, based on previous experience by Lo et al. in reference 32.

712. c.

For two alleles, the prior probability is $1/2$ for "did inherit" and $1/2$ for "did not inherit." The conditional probability is $2/3$ for "did inherit" and 1 for "did not inherit." The joint probability is $2/6$ for "did inherit" and $1/2$ for "did not inherit." Thus, the posterior probability of inheriting, knowing that he is asymptomatic at age 40, is $[(2/6) \div (2/6 + 1/2)]$ or $(2/6) \div (5/6)$ or 40%.

713. e.

Even though proteins start with methionine (AUG) in the primary transcript, the mRNA itself is capped with 7-methylguanosine in a unique 5′-5′ bond. Answers *c* and *d* are incorrect because there is no repetitive "leader" sequence in mRNA, which does, however, have a poly-A tail.

714. 0.040 or 4.0%

The Hardy-Weinburg law is $p^2 + 2pq + q^2 = 1$ where $p^2$ is the frequency of non-carriers, 2pq is the frequency of heterozygote carriers, and $q^2$ is the frequency of homozygous affected individuals. In this case, $q^2 = 1/2500$ or 0.0004 and, thus, $q = 0.002$. Assuming that $p = 1$ because the disease is so rare, $2pq = 0.002$ or, expressed as a percentage, 4.0%. If p is calculated from the formula $p + q = 1$, the answer is 3.992% or, to two significant figures, still 4.0%.

715. d.

See Chapter 3 from the reference. Since histones are completely replaced, answers *a* and *b* are incorrect. Neither fibronectin nor cumulin is a part of the sperm's genetic apparatus.

716. b.

The other mutations listed are associated with "ABC" genes that are not *ABCA3*. *ABCA1* is associated with Tangier disease. *ABCG5* and *ABCG8*, not *ABCA3*, are expressed in the liver and intestine. *ABCA4* is involved in retinal mutations.

717. e.

From the paper cited, all of the mutations listed are associated with DIP and/or PAP.

718. True.

According to the paper cited, this is a true statement.

719. c.

Of the mutations (no polymorphisms were listed in the question), only the 4552insT mutation causes a frameshift, although there are other mutations not listed in the question that can cause a frameshift.

720. d.

According to Chapter 7 of the reference, all three answers are valid explanations for the "metabolic safety net" that allows cells to handle toxic insults with primary and secondary backup systems.

721. d.

HCDA shows all three SNPs, including the G208A variant discovered in the study cited.

722. c.

The A79C mutation results in $Lys^{27}$>Gln and G208A results in $Ala^{70}$>Thr, while the T435C mutation yields $Thr^{145}$>Thr.

723. True

According to the authors and the data on the reduced enzyme activity, this statement is true, even though no disease state from this mutation has resulted to date.

724. b.

Both $TPMT^{L}$ and $TPMT^{H}$ are coded by a single allele. Thus, answer *b* is correct because a single allele leads to a trimodal distribution with the activity expressions given. Answer *c* is incorrect because the frequencies are reversed. Answer *d* is incorrect because the most common of the eight polymorphisms is TPMT*3, which differs from the wild-type (TPMT*1) at residues 154 (A→T) and 240 (Y→C).

725. c.

In Table 7-3, $X_i$ is not even mentioned, and the other answers are associated with Prader-Willi, Smith-Magenis, and Williams syndromes.

726. $1.71 \times 10^{-4}$

For an X-linked recessive disease, $\mu = sq/3$. Thus, $\mu = [(0.023)(0.02236)]/3$. ($q^2 = 5 \times 10^{-4}$ and, therefore, $q = 0.02236$)

727. e.

All of the markers were those used in the reference.

728. e.

According to the reference, CGH cannot detect any of the listed mutations.

729. b.

That is the technique used in the reference, and the other four answers were just made up.

730. a.

The correct answer is given in the box on page 98 of the reference. Answer *b* is the definition of an inversion. Answer *c* is the definition of a reciprocal translocation. Answer *d* is incorrect because the word "metacentric" has been inserted. Answer *e* is incorrect because it is the reverse of the definition of a Robertsonian translocation.

731. b.

Answers *a*, *c*, and *d* are opposites of the statements given in the reference and, thus, *e* is also not correct.

732. e.

All of the statements are true, according to the reference.

733. a.

This formula is given in the reference. All of the other answers are just concoctions.

734. e.

These are all conclusions reached by the authors.

735. True

Please see the box on page 48 of the reference.

736. e.

This answer is taken from a composite of the references.

737. e.

This answer was derived from the Bullinger et al. reference.

738. d.

Genetic signatures could be obtained for all other mutations listed. Trisomy 8 is, from the reference and previous publications, too diverse to obtain a single signature.

739. e.

The correct answer is *e* because t(15;17) is associated with all of the other four choices and their general category.

740. True.

According to the Bullinger reference, the statement is true, as this is what is proposed.

741. True.

Please see the Valk et al. reference.

742. False

The statement is the opposite of the one given in Figure 5-1 of the reference.

743. c.

Please see the reference. Answers *a* and *b* refer to the TNF receptor. Answer *d* is the opposite of the statement in the reference and answer *e* refers to the coding region rather than the promoter region.

744. c.

Answers *a* and *b* resolve FRET assay loss of fluorescence issues that are not related to the 3′-phosphate cap. Thus, answers *d* and *e* also cannot be correct.

745. True

The statement agrees with Figure 6-12 in the reference.

746. e.

Please see the reference.

747. d.

The authors used this as the correct answer in the reference.

748. True

According to the reference, bone marrow specimens are preferred for the reason given in the question (except for M3v).

749. d.

Answers *a*, *b*, and *c* were recommended by the authors in the reference.

750. e.

According to the reference, there is no association between the C282Y allele and the risk of developing preeclampsia. Since the basis of the hypothesis was iron overload and the C282Y allele, presumably the same applies to the H63D allele. Thus, the correct answer is *e*.

# References

Note to the reader: the numbers on the left refer to the questions. The references listed on the right are the sources of the answers to those questions.


1–3. Lewin B. Genes VII. Oxford, UK: Oxford University Press, 2000.

4. Sambrook J, Russell DW. Molecular cloning: A laboratory manual, 3rd ed. Cold Spring Harbor, NY: Cold Spring Harbor Press, 2001.

5–8. Lewin, B. Genes VII. Oxford, UK: Oxford University Press, 2000.

9. Sambrook J, Russell DW. Molecular cloning: A laboratory manual, 3rd ed. Cold Spring Harbor, NY: Cold Spring Harbor Press, 2001.

10. Esch RK. Basic nucleic acid procedures. In: Coleman WB, Tsongalis GJ, eds. Molecular diagnostics for the clinical laboratorian. Totowa, NJ: Humana Press, 2000.

11–12. Sambrook J, Russell DW. Molecular cloning: A laboratory manual, 3rd ed. Cold Spring Harbor, NY: Cold Spring Harbor Press, 2001.

13. Coleman WB. An overview of nucleic acid chemistry, structure, and function: The foundations of molecular biology. In: Coleman WB, Tsongalis GJ, eds. Molecular diagnostics for the clinical laboratorian. Totowa, NJ: Humana Press, 2000.

14. Tsongalis GJ. An overview of molecular genetics. In: Coleman WB, Tsongalis GJ, eds. Molecular diagnostics for the clinical laboratorian. Totowa, NJ: Humana Press, 2000.

15–19. Sambrook J, Russell DW. Molecular cloning: A laboratory manual, 3rd ed. Cold Spring Harbor, NY: Cold Spring Harbor Press, 2001.

20–21. Esch RK. Basic nucleic acid procedures. In: Coleman WB, Tsongalis GJ, eds. Molecular diagnostics for the clinical laboratorian. Totowa, NJ: Humana Press, 2000.

22–30. Sambrook J, Russell DW. Molecular cloning: A laboratory manual, 3rd ed. Cold Spring Harbor, NY: Cold Spring Harbor Press, 2001.

31–33. Sorscher DH. DNA amplication techniques: An overview. In: Coleman WB, Tsongalis GJ, eds. Molecular diagnostics for the clinical laboratorian. Totowa, NJ: Humana Press, 2000.

34. Sambrook J, Russell DW. Molecular cloning: A laboratory manual, 3rd ed. Cold Spring Harbor, NY: Cold Spring Harbor Press, 2001.

35–36. Sorscher DH. DNA amplication techniques: An overview. In: Coleman WB, Tsongalis GJ, eds. Molecular diagnostics for the clinical laboratorian. Totowa, NJ: Humana Press, 2000.

37–51. Sambrook J, Russell DW. Molecular cloning: A laboratory manual, 3rd ed. Cold Spring Harbor, NY: Cold Spring Harbor Press, 2001.

52. Ausubel FM, Brett R, Kingston RE, Moore DD, Seidman JG, Smith JA, et al., eds. Short protocols in molecular biology, 4th ed. New York: Wiley, 1999.

53. Strachen T, Reed AP, eds. Human molecular genetics 3. London: Garland Science Taylor & Francis Group, 2004.

54. Esch RK. Basic nucleic acid procedures. In: Coleman WB, Tsongalis GJ, eds. Molecular diagnostics for the clinical laboratorian. Totowa, NJ: Humana Press, 2000.

55–64. Watson JD, Hopkins NH, Roberts JW, Steitz JA, Weiner AM, eds. Molecular biology of the gene, 4th ed. Menlo Park, CA: Benjamin/Cummings, 1987.

Budavari S, ed. The Merck index, 13th ed. Rahway, NJ: Merck & Co., Inc., 2002.

65. Fearon ER. Tumor suppressor genes. In: Vogelstein B, Kinzler KW, eds. The genetic basis of human cancer, 2nd ed. New York: McGraw-Hill Publishers, 2002.

66. Latchman DS. Eukaryotic transcription factors. Biochem J 1990;270: 281–289.

Rhodes D and Klug A. Zinc fingers. Sci Am 1993;Feb:56–65.

Klug A. Zinc finger peptides for the regulation of gene expression. J Mol Biol 1999;293:215–218.

67. Park M. Oncogenes. In: Vogelstein B, Kinzler KW, eds. The genetic basis of human cancer, 2nd ed. New York: McGraw-Hill Publishers, 2002.

68–89. Williams S, Slatko B, et al. New England Biolabs/Smith College Molecular biology summer workshops. [Notes]. Northampton, MA, June–July 2003.

Tsongalis GJ, Coleman WB. Molecular technologies. In: Tsongalis GJ, Coleman WB, eds. Molecular diagnostics: a training and study guide. Washington, DC: AACC Press, 2002.

New England Biolabs. 2002–03 Catalog and technical reference. Beverly, MA: New England BioLabs, Inc., 2002.

Watson JD, Hopkins NH, Roberts JW, Steitz JA, Weiner AM, eds. Molecular biology of the gene, 4th ed. Menlo Park, CA: Benjamin/Cummings, 1987.

90. Bootsma D, Kraemer KH, Cleaver JE, and Hoeijmakers JHJ. Nucleotide excision repair syndromes: Xeroderma pigmentosum, Cockayne syndrome, and trichothiodystrophy. In: Vogelstein B, Kinzler KW, eds. The genetic basis of human cancer, 2nd ed. New York: McGraw-Hill Publishers, 2002.

91. Beutler E. Genetic principles and molecular biology. In: Beutler E, Lichtman MA, Coller BS, Kipps TJ, and Seligsohn U, eds. Williams hematology, 6th ed. New York: McGraw-Hill Medical Publishing Division, 2001.

92. Ross DW. Introduction to molecular medicine, 3rd ed. New York: Springer-Verlag, 2002.

93. Aaltonen LA. Peutz-Jehgers syndrome. In: Vogelstein B, Kinzler KW, eds. The genetic basis of human cancer, 2nd ed. New York: McGraw-Hill Publishers, 2002.

94. Auerbach AD, Buchwald M, Joenje H. Fanconi anemia. In: Vogelstein B, Kinzler KW, eds. The genetic basis of human cancer, 2nd ed. New York: McGraw-Hill Publishers, 2002.

95. Ferreira-Gonzalez A, Wilkinson DS, and Garrett CT. Introduction to molecular diagnostics. In: McClatchey KD, ed. Clinical laboratory medicine, 2nd ed. Philadelphia: Lippincott Williams & Wilkins, 2002.

96–99. Sandin RL. Molecular biology of infectious diseases. In: McClatchey KD, ed. Clinical laboratory medicine, 2nd ed. Philadelphia: Lippincott Williams & Wilkins, 2002.

100–101. Rohifs EM and Highsmith WE. PCR-based methods for mutation detection. In: Coleman WB, Tsongalis GJ, eds. Molecular diagnostics for the clinical laboratorian. Totowa, NJ: Humana Press, 2000.

102–103. Sambrook J, Russell DW. Molecular cloning: A laboratory manual, 3rd ed. Cold Spring Harbor, NY: Cold Spring Harbor Press, 2001.

Anthony DC, Montine TJ, Valentine WM, and Graham DG. Toxic responses of the nervous system. In: Klaassen CD, ed. Casarett & Doull's toxicology. The basic science of poisons, 6th ed. New York: McGraw-Hill Medical Publishing Division, 2001.

104–110. Rohifs EM, Highsmith WE. PCR-based methods for mutation detection. In: Coleman WB, Tsongalis GJ, eds. Molecular diagnostics for the clinical laboratorian. Totowa, NJ: Humana Press, 2000.

111–112. Esch RK. Basic nucleic acid procedures. In: Coleman WB, Tsongalis GJ, eds. Molecular diagnostics for the clinical laboratorian. Totowa, NJ: Humana Press, 2000.

113–115. Sambrook J, Russell DW. Molecular cloning: A laboratory manual, 3rd ed. Cold Spring Harbor, NY: Cold Spring Harbor Press, 2001.

116. Esch RK. Basic nucleic acid procedures. In: Coleman WB, Tsongalis GJ, eds. Molecular diagnostics for the clinical laboratorian. Totowa, NJ: Humana Press, 2000.

117. Sambrook J, Russell DW. Molecular cloning: A laboratory manual, 3rd ed. Cold Spring Harbor, NY: Cold Spring Harbor Press, 2001.

118. Brown TA. Genomes, 2nd ed. New York: Wiley-Liss, 2002.

119–120. Esch RK. Basic nucleic acid procedures. In: Coleman WB, Tsongalis GJ, eds. Molecular diagnostics for the clinical laboratorian. Totowa, NJ: Humana Press, 2000.

121–130. Sambrook J, Russell DW. Molecular cloning: A laboratory manual, 3rd ed. Cold Spring Harbor, NY: Cold Spring Harbor Press, 2001.

131–132. Cowlen MS. Nucleic acid hybridization and amplification in situ: Principles and applications in molecular pathology. In: Coleman WB, Tsongalis GJ, eds. Molecular diagnostics for the clinical laboratorian. Totowa, NJ: Humana Press, 2000.

133–135. Tsongalis GJ. An overview of molecular genetics. In: Coleman WB, Tsongalis GJ, eds. Molecular diagnostics for the clinical laboratorian. Totowa, NJ: Humana Press, 2000.

136–171. Williams S, Slatko B, et al. New England Biolabs/Smith College Molecular biology summer workshops. [Notes]. Northampton, MA, June–July 2003.

New England Biolabs. 2002–03 Catalog and technical reference. Beverly, MA: New England BioLabs, Inc., 2002.

172. Bootsma D, Kraemer KH, Cleaver JE, Hoeijmakers JHJ. Nucleotide excision repair syndromes: Xeroderma pigmentosum, Cockayne syndrome, and trichothiodystrophy. In: Vogelstein B, Kinzler KW, eds. The genetic basis of human cancer, 2nd ed. New York: McGraw-Hill Publishers, 2002.

173. Schwartz S, Palmer CG. Basic cytogenetics. In: McClatchey KD, ed. Clinical laboratory medicine, 2nd ed. Philadelphia: Lippincott Williams & Wilkins, 2002.

174. Van Dyke DL, Wiktor A. Clinical cytogenetics. In: McClatchey KD, ed. Clinical laboratory medicine, 2nd ed. Philadelphia: Lippincott Williams & Wilkins, 2002.

Feinberg AP. Genomic imprinting and cancer. In: Vogelstein B, Kinzler KW, eds. The genetic basis of human cancer, 2nd ed. New York: McGraw-Hill Publishers, 2002.

175–178. Malkin D. Li-Fraumeni syndrome. In: Vogelstein B, Kinzler KW, eds. The genetic basis of human cancer, 2nd ed. New York: McGraw-Hill Publishers, 2002.

179. Schwartz S, Palmer CG. Basic cytogenetics. In: McClatchey KD, ed. Clinical laboratory medicine, 2nd ed. Philadelphia: Lippincott Williams & Wilkins, 2002.

180–182. Boland CR. Hereditary nonpolyposis colorectal cancer (HNPCC). In: Vogelstein B, Kinzler KW, eds. The genetic basis of human cancer, 2nd ed. New York: McGraw-Hill Publishers, 2002.

183. Schwartz S, Palmer CG. Basic cytogenetics. In: McClatchey KD, ed. Clinical laboratory medicine, 2nd ed. Philadelphia: Lippincott Williams & Wilkins, 2002.

Zhang Y-H, Zhang M. A dictionary of gene technology terms. New York: Parthenon Publishing Group, 2001.

184. Alberts B, Johnson A, Lewis J, Raff M, Roberts K, Walter P. How cells read the genome: From DNA to protein. In: Molecular biology of the cell, 4th ed. New York: Garland Science Taylor & Francis Group, 2002.

185–191. Newsham IF, Hadjistilianou T, Cavenee WK. Retinoblastoma. In: Vogelstein B, Kinzler KW, eds. The genetic basis of human cancer, 2nd ed. New York: McGraw-Hill Publishers, 2002.

192–196. Gutmann DH, Collins FS. Neurofibromatosis 1. In: Vogelstein B, Kinzler KW, eds. The genetic basis of human cancer, 2nd ed. New York: McGraw-Hill Publishers, 2002.

MacCollin M, Gusella J. Neurofibromatosis 2. In: Vogelstein B, Kinzler KW, eds. The genetic basis of human cancer, 2nd ed. New York: McGraw-Hill Publishers, 2002.

197. Linehan WM, Zhar B, Klausner RD. Renal carcinoma. In: Vogelstein B, Kinzler KW, eds. The genetic basis of human cancer, 2nd ed. New York: McGraw-Hill Publishers, 2002.

198. Kant JA. Molecular biology of inherited diseases. In: McClatchey KD, ed. Clinical laboratory medicine, 2nd ed. Philadelphia: Lippincott Williams & Wilkins, 2002.

199. Ross DW. Introduction to molecular medicine, 3rd ed. New York: Springer-Verlag, 2002.

200. Cooper DN, Krawczak M, Antonarakis SE. The nature and mechanisms of human gene mutation. In: Vogelstein B, Kinzler KW, eds. The genetic basis of human cancer, 2nd ed. New York: McGraw-Hill Publishers, 2002.

201–211. Ross DW. Introduction to molecular medicine, 3rd ed. New York: Springer-Verlag, 2002.

212. Sandin RL. Molecular biology of infectious diseases. In: McClatchey KD, ed. Clinical laboratory medicine, 2nd ed. Philadelphia: Lippincott Williams & Wilkins, 2002.

213–214. Esch RK. Basic nucleic acid procedures. In: Coleman WB, Tsongalis GJ, eds. Molecular diagnostics for the clinical laboratorian. Totowa, NJ: Humana Press, 2000.

215. Sambrook J, Russell DW. Molecular cloning: A laboratory manual, 3rd ed. Cold Spring Harbor, NY: Cold Spring Harbor Press, 2001.

216–217. Esch RK. Basic nucleic acid procedures. In: Coleman WB, Tsongalis GJ, eds. Molecular diagnostics for the clinical laboratorian. Totowa, NJ: Humana Press, 2000.

218–225. Sambrook J, Russell DW. Molecular cloning: A laboratory manual, 3rd ed. Cold Spring Harbor, NY: Cold Spring Harbor Press, 2001.

226. Sorscher DH. DNA amplication techniques: An overview. In: Coleman WB, Tsongalis GJ, eds. Molecular diagnostics for the clinical laboratorian. Totowa, NJ: Humana Press, 2000.

227–228. Sambrook J, Russell DW. Molecular cloning: A laboratory manual, 3rd ed. Cold Spring Harbor, NY: Cold Spring Harbor Press, 2001.

229. Sorscher DH. DNA amplication techniques: An overview. In: Coleman WB, Tsongalis GJ, eds. Molecular diagnostics for the clinical laboratorian. Totowa, NJ: Humana Press, 2000.

230. Lewin B. Genes VII. Oxford, UK: Oxford University Press, 2000.

231. Sambrook J, Russell DW. Molecular cloning: A laboratory manual, 3rd ed. Cold Spring Harbor, NY: Cold Spring Harbor Press, 2001.

232. Lewin B. Genes VII. Oxford, UK: Oxford University Press, 2000.

233–234. Sambrook J, Russell DW. Molecular cloning: A laboratory manual, 3rd ed. Cold Spring Harbor, NY: Cold Spring Harbor Press, 2001.

235. Lewin B. Genes VII. Oxford, UK: Oxford University Press, 2000.

236–245. Sambrook J, Russell DW. Molecular cloning: A laboratory manual, 3rd ed. Cold Spring Harbor, NY: Cold Spring Harbor Press, 2001.

246–248. Sorscher DH. DNA amplication techniques: An overview. In: Coleman WB, Tsongalis GJ, eds. Molecular diagnostics for the clinical laboratorian. Totowa, NJ: Humana Press, 2000.

249–251. Sambrook J, Russell DW. Molecular cloning: A laboratory manual, 3rd ed. Cold Spring Harbor, NY: Cold Spring Harbor Press, 2001.

252–261. Park M. Oncogenes. In: Vogelstein B, Kinzler KW, eds. The genetic basis of human cancer, 2nd ed. New York: McGraw-Hill Publishers, 2002.

262–280. Williams S, Slatko B, et al. New England Biolabs/Smith College Molecular biology summer workshops. [Notes]. Northampton, MA, June–July 2003.

281. Gatti RA. Ataxia-telangiectasia. In: Vogelstein B, Kinzler KW, eds. The genetic basis of human cancer, 2nd ed. New York: McGraw-Hill Publishers, 2002.

282–289. Fairbanks VF, Klee GG. Biochemical aspects of hematology. In: Burtis CA, Ashwood ER. Tietz textbook of clinical chemistry, 3rd ed. Philadelphia: W B Saunders Company, 1999.

Maniatis T. Molecular genetics and biosynthesis of hemoglobin. In: Bunn HF, Forget BG, eds. Hemoglobin: Molecular, genetic, and clinical aspects. Philadelphia: W B Saunders Company, 1986.

290. German J and Ellis NA. Bloom syndrome. In: Vogelstein B, Kinzler KW, eds. The genetic basis of human cancer, 2nd ed. New York: McGraw-Hill Publishers, 2002.

291–296. Couch FJ and Weber BL. Breast cancer. In: Vogelstein B, Kinzler KW, eds. The genetic basis of human cancer, 2nd ed. New York: McGraw-Hill Publishers, 2002.

297. Cho KR. Cervical cancer. In: Vogelstein B, Kinzler KW, eds. The genetic basis of human cancer, 2nd ed. New York: McGraw-Hill Publishers, 2002.

298–299. Ross DW. Introduction to molecular medicine, 3rd ed. New York: Springer-Verlag, 2002.

300–302. Schellenburg GD, Miki T, Yu C-E, and Nakura J. Werner syndrome. In: Vogelstein B, Kinzler KW, eds. The genetic basis of human cancer, 2nd ed. New York: McGraw-Hill Publishers, 2002.

303–324. Lewin B. Genes VII. Oxford, UK: Oxford University Press, 2000.

325–329. Haber DA. Wilms tumor. In: Vogelstein B, Kinzler KW, eds. The genetic basis of human cancer, 2nd ed. New York: McGraw-Hill Publishers, 2002.

330–334. Schwartz S, Palmer CG. Basic cytogenetics. In: McClatchey KD, ed. Clinical laboratory medicine, 2nd ed. Philadelphia: Lippincott Williams & Wilkins, 2002.

335–343. Alberts B, Johnson A, Lewis J, Raff M, Roberts K, Walter P. Germ cells and fertilization. In: Molecular biology of the cell, 4th ed. New York: Garland Science Taylor & Francis Group, 2002.

344–345. Lewin B. Genes VII. Oxford, UK: Oxford University Press, 2000.

346. Schwartz S, Palmer CG. Basic cytogenetics. In: McClatchey KD, ed. Clinical laboratory medicine, 2nd ed. Philadelphia: Lippincott Williams & Wilkins, 2002.

347. German J and Ellis NA. Bloom syndrome. In: Vogelstein B, Kinzler KW, eds. The genetic basis of human cancer, 2nd ed. New York: McGraw-Hill Publishers, 2002.

348. Kant JA. Molecular biology of inherited diseases. In: McClatchey KD, ed. Clinical laboratory medicine, 2nd ed. Philadelphia: Lippincott Williams & Wilkins, 2002.

349. Ross DW. Introduction to molecular medicine, 3rd ed. New York: Springer-Verlag, 2002.

350–363. Lichtman MA. Classification and clinical manifestations of the clonal myeloid disorders. In: Beutler E, Lichtman MA, Coller BS, Kipps TJ, and Seligsohn U, eds. Williams hematology, 6th ed. New York: McGraw-Hill Medical Publishing Division, 2001.

Lichtman MA, Brennan JK. Myelopdysplasitc disorders (indolent clonal myeloid diseases and oligoblastic leukemia). In: Beutler E, Lichtman MA, Coller BS, Kipps TJ, Seligsohn U, eds. Williams hematology, 6th ed. New York: McGraw-Hill Medical Publishing Division, 2001.

Lichtman MA, Liesveld JL. Acute myelogenous leukemia. In: Beutler E, Lichtman MA, Coller BS, Kipps TJ, Seligsohn U, eds. Williams hematology, 6th ed. New York: McGraw-Hill Medical Publishing Division, 2001.

Lichtman MA, Liesveld JL. Chronic myelogenous leukemia and related disorders. In: Beutler E, Lichtman MA, Coller BS, Kipps TJ, and Seligsohn U, eds. Williams hematology, 6th ed. New York: McGraw-Hill Medical Publishing Division, 2001.

Kipps TJ. Clasification of malignant lymphoid disorders. In: Beutler E, Lichtman MA, Coller BS, Kipps TJ, and Seligsohn U, eds. Williams hematology, 6th ed. New York: McGraw-Hill Medical Publishing Division, 2001.

Pui C-H. Acute lymphoblastic leukemia. In: Beutler E, Lichtman MA, Coller BS, Kipps TJ, and Seligsohn U, eds. Williams hematology, 6th ed. New York: McGraw-Hill Medical Publishing Division, 2001.

Kipps TJ. Chronic lymphocytic leukemia and related disorders. In: Beutler E, Lichtman MA, Coller BS, Kipps TJ, and Seligsohn U, eds. Williams hematology, 6th ed. New York: McGraw-Hill Medical Publishing Division, 2001.

Longo DL. Genetic abnormalities in lymphoid malignancies. In: Vogelstein B, Kinzler KW, eds. The genetic basis of human cancer, 2nd ed. New York: McGraw-Hill Publishers, 2002.

Cooper DN, Krawczak M, Antonarakis SE. The nature and mechanisms of human gene mutation. In: Vogelstein B, Kinzler KW, eds. The genetic basis of human cancer, 2nd ed. New York: McGraw-Hill Publishers, 2002.

364–366. Kinzler KW, Vogelstein B. Colorectal tumors. In: Vogelstein B, Kinzler KW, eds. The genetic basis of human cancer, 2nd ed. New York: McGraw-Hill Publishers, 2002.

367–389. Williams S, Slatko B, et al. New England Biolabs/Smith College Molecular biology summer workshops. [Notes]. Northampton, MA, June–July 2003.

New England Biolabs. 2002–03 Catalog and technical reference. Beverly, MA: New England BioLabs, Inc., 2002.

Tsongalis GJ, Coleman WB. Molecular technologies. In: Tsongalis GJ, Coleman WB, eds. Molecular diagnostics. A training and study guide. Washington, DC: AACC Press, 2002.

390–391. Gatti RA. Ataxia-telangiectasia. In: Vogelstein B, Kinzler KW, eds. The genetic basis of human cancer, 2nd ed. New York: McGraw-Hill Publishers, 2002.

392. Ferreira-Gonzalez A, Wilkinson DS, Garrett CT. Introduction to molecular diagnostics. In: McClatchey KD, ed. Clinical laboratory medicine, 2nd ed. Philadelphia: Lippincott Williams & Wilkins, 2002.

393–394. Lewin B. Genes VII. Oxford, UK: Oxford University Press, 2000.

395. Howe JR. Juvenile polyposis syndrome. In: Vogelstein B, Kinzler KW, eds. The genetic basis of human cancer, 2nd ed. New York: McGraw-Hill Publishers, 2002.

396. Sambrook J, Russell DW. Molecular cloning: A laboratory manual, 3rd ed. Cold Spring Harbor, NY: Cold Spring Harbor Press, 2001.

397–400. Sandin RL. Molecular biology of infectious diseases. In: McClatchey KD, ed. Clinical laboratory medicine, 2nd ed. Philadelphia: Lippincott Williams & Wilkins, 2002.

Unger ER, Piper MA. Nucleic acid biochemistry and diagnostic applications. In: Burtis CA, Ashwood ER. Tietz textbook of clinical chemistry, 3rd ed. Philadelphia: W B Saunders Company, 1999.

401. Sambrook J, Russell DW. Molecular cloning: A laboratory manual, 3rd ed. Cold Spring Harbor, NY: Cold Spring Harbor Press, 2001.

402. Sorscher DH. DNA amplication techniques: An overview. In: Coleman WB, Tsongalis GJ, eds. Molecular diagnostics for the clinical laboratorian. Totowa, NJ: Humana Press, 2000.

403–405. Sambrook J, Russell DW. Molecular cloning: A laboratory manual, 3rd ed. Cold Spring Harbor, NY: Cold Spring Harbor Press, 2001.

406. Digene Web site. http://www.digene.com (Accessed April 8, 2004).

407–408. Sorscher DH. DNA amplication techniques: An overview. In: Coleman WB, Tsongalis GJ, eds. Molecular diagnostics for the clinical laboratorian. Totowa, NJ: Humana Press, 2000.

409. Sambrook J, Russell DW. Molecular cloning: A laboratory manual, 3rd ed. Cold Spring Harbor, NY: Cold Spring Harbor Press, 2001.

410. Sorscher DH. DNA amplication techniques: An overview. In: Coleman WB, Tsongalis GJ, eds. Molecular diagnostics for the clinical laboratorian. Totowa, NJ: Humana Press, 2000.

411. Sambrook J, Russell DW. Molecular cloning: A laboratory manual, 3rd ed. Cold Spring Harbor, NY: Cold Spring Harbor Press, 2001.

412. Fromm MF, Eichelbaum M. The pharmacogenomics of human p-glycoprotein. In: Licinio J, Wong M-L, eds. Pharmacogenomics. The search for individualized therapies. Weinbaum, Germany: Wiley-VCH, 2003.

413. Strachan T, Reed AP, eds. Human molecular genetics 3. London: Garland Science Taylor & Francis Group, 2004.

414–415. Sambrook J, Russell DW. Molecular cloning: A laboratory manual, 3rd ed. Cold Spring Harbor, NY: Cold Spring Harbor Press, 2001.

416. Hurley TD, Edenburg HJ, Li T-K. Pharmacogenomics of alcoholism. In: Licinio J, Wong M-L, eds. Pharmacogenomics. The search for individualized therapies. Weinbaum, Germany: Wiley-VCH, 2003.

417–419. Sambrook J, Russell DW. Molecular cloning: A laboratory manual, 3rd ed. Cold Spring Harbor, NY: Cold Spring Harbor Press, 2001.

420. Innocenti F, Iyer L, Ratain MJ. Pharmacogenomics of chemotherapeutic agents in cancer treatment. In: Licinio J, Wong M-L, eds. Pharmacogenomics. The search for individualized therapies. Weinbaum, Germany: Wiley-VCH, 2003.

421. Sorscher DH. DNA amplication techniques: An overview. In: Coleman WB, Tsongalis GJ, eds. Molecular diagnostics for the clinical laboratorian. Totowa, NJ: Humana Press, 2000.

422–424. Rohifs EM, Highsmith WE. PCR-based methods for mutation detection. In: Coleman WB, Tsongalis GJ, eds. Molecular diagnostics for the clinical laboratorian. Totowa, NJ: Humana Press, 2000.

425. Wick MJ, Crifasi PA, Wang Z, Thibodeau SN. Genetic basis of neurologic and neuromuscular diseases. In: Coleman WB, Tsongalis GJ, eds. Molecular diagnostics for the clinical laboratorian. Totowa, NJ: Humana Press, 2000.

426–427. Rohifs EM, Highsmith WE. PCR-based methods for mutation detection. In: Coleman WB, Tsongalis GJ, eds. Molecular diagnostics for the clinical laboratorian. Totowa, NJ: Humana Press, 2000.

428–429. Sambrook J, Russell DW. Molecular cloning: A laboratory manual, 3rd ed. Cold Spring Harbor, NY: Cold Spring Harbor Press, 2001.

430. Lewin B. Genes VII. Oxford, UK: Oxford University Press, 2000.

431. Sambrook J, Russell DW. Molecular cloning: A laboratory manual, 3rd ed. Cold Spring Harbor, NY: Cold Spring Harbor Press, 2001.

432. Lewin B. Genes VII. Oxford, UK: Oxford University Press, 2000.

433–434. Sambrook J, Russell DW. Molecular cloning: A laboratory manual, 3rd ed. Cold Spring Harbor, NY: Cold Spring Harbor Press, 2001.

435. Lewin B. Genes VII. Oxford, UK: Oxford University Press, 2000.

436–438. Sambrook J, Russell DW. Molecular cloning: A laboratory manual, 3rd ed. Cold Spring Harbor, NY: Cold Spring Harbor Press, 2001.

439–478. Lewin B. Genes VII. Oxford, UK: Oxford University Press, 2000.

479. Alberts B, Johnson A, Lewis J, Raff M, Roberts K, Walter P. How cells read the genome: From DNA to protein. In: Molecular biology of the cell, 4th ed. New York: Garland Science Taylor & Francis Group, 2002.

480–488. Lewin B. Genes VII. Oxford, UK: Oxford University Press, 2000.

489. Sambrook J, Russell DW. Molecular cloning: A laboratory manual, 3rd ed. Cold Spring Harbor, NY: Cold Spring Harbor Press, 2001.

490. Lewin B. Genes VII. Oxford, UK: Oxford University Press, 2000.

491–496. Sambrook J, Russell DW. Molecular cloning: A laboratory manual, 3rd ed. Cold Spring Harbor, NY: Cold Spring Harbor Press, 2001.

497. Rohifs EM, Highsmith WE. PCR-based methods for mutation detection. In: Coleman WB, Tsongalis GJ, eds. Molecular diagnostics for the clinical laboratorian. Totowa, NJ: Humana Press, 2000.

498. Lewin B. Genes VII. Oxford, UK: Oxford University Press, 2000.

499. Tsongalis GJ. An overview of molecular genetics. In: Coleman WB, Tsongalis GJ, eds. Molecular diagnostics for the clinical laboratorian. Totowa, NJ: Humana Press, 2000.

500. Brown TA. Genomes, 2nd ed. New York: Wiley-Liss, 2002.

501–505. Schwarzacher T, Heslop-Harrison P. Practical in situ hybridization. Oxfordshire, UK: Bios Scientific Publishers, 1999.

506–513. Hannon GJ. RNAi: A guide to gene silencing. Cold Spring Harbor, NY: Cold Spring Harbor Laboratory Press, 2003.

514–517. Ekenberg S, Hudson G. RNase protection assay system: A versatile technique for the analysis of RNA. Promega Notes Magazine 1994;46:14–17.

518–522. Butler JM. Forensic DNA typing. San Diego, CA: Academic Press, 2001.

523. Strachan T, Read AP. Human molecular genetics 2, 2nd ed. Oxford, UK: Bios Scientific Publishers, Ltd., 1999.

524. Merscher S, Hanselmann R, Welter C, Dooley S. Nuclear runoff transcription analysis using chemiluminescent detection. Biotechniques 1994;16: 1024–1026.

525–533. Knudsen S. Guide to analysis of DNA microarray data, 2nd ed. New York: John Wiley & Sons, 2004.

534–536. Strachan T, Read AP. Human molecular genetics 2, 2nd ed. Oxford, UK: Bios Scientific Publishers, Ltd., 1999.

537–543. Decker J, Reischl U. Molecular diagnosis of infectious diseases, 2nd ed. Totowa, NJ: Humana Press, 2003.

544–549. National Library of Medicine. The NCBI Handbook. Bethesda, MD: National Center for Biotechnology Information, 2002.

550–554. Williams S, Slatko B, et al. New England Biolabs/Smith College Molecular biology summer workshops. [Notes]. Northampton, MA, June–July 2003.

555–564. Brown TA. Genomes, 2nd ed. New York: Wiley-Liss, 2002.

565–566. Lewin B. Genes VII. Oxford, UK: Oxford University Press, 2000.

567–607. Griffiths AJ, Miller JH, Suzuki DT, Lewontin RC, Gelbart WM. Introduction to genetic analysis, 7th ed. New York: W.H. Freeman & Co., 1999.

608–613. Walsh G. Biopharmaceuticals: Biochemistry and biotechnology, 2nd ed. Chichester, UK: John Wiley & Sons, Ltd., 2003.

614–615. Strachan T, Read AP. Human molecular genetics 2, 2nd ed. Oxford, UK: Bios Scientific Publishers, Ltd., 1999.

616. Koonin EV, Galperin MY. Sequence—evolution—function: Computational approaches in comparative genomics. Norwell, MA: Kluwer Academic Publishers, 2003.

617. Salyers AA, Whitt DD. Bacterial pathogenesis: A molecular approach. Washington, DC: ASM Press, 1994.

618–619. Tortora GJ, Funke BR, Case CL. Microbiology: An introduction, 8th ed. San Francisco, CA: Pearson Education, Inc., 2004.

620. Salyers AA, Whitt DD. Bacterial pathogenesis: A molecular approach. Washington, DC: ASM Press, 1994.

621–651. Tortora GJ, Funke BR, Case CL. Microbiology: An introduction, 8th ed. San Francisco, CA: Pearson Education, Inc., 2004.

652–681. Prescott LM, Harley JP, Klein DA. Microbiology, 5th ed. Boston, MA: McGraw-Hill, 2002.

682–696. Bauman RW. Microbiology. San Francisco, CA: Pearson Education, Inc., 2004.

697–700. Prescott LM, Harley JP, Klein DA. Microbiology, 5th ed. Boston, MA: McGraw-Hill, 2002.

701. Horwitz M. Basic concepts in medical genetics. A student's survival guide. New York: The McGraw-Hill Companies, Inc., 2000.

702–711. Nasis O, Thompson S, Hong T, Sherwood M, Radcliffe S, Jackson L, et al. Improvement in sensitivity of allele-specific PCR facilitates reliable noninvasive prenatal detection of cystic fibrosis. Clin Chem 2004;50(4):694–701.

712. Horwitz M. Basic concepts in medical genetics. A student's survival guide. New York: The McGraw-Hill Companies, Inc., 2000.

713. Alberts B, Johnson A, Lewis J, Raff M, Roberts K, Walter P. DNA replication, repair, and recombination. In: Molecular biology of the cell, 4th ed. New York: Garland Science Taylor & Francis Group, 2002.

714. Tsongalis GJ, Coleman WB. Molecular diagnostics. A training and study guide. Washington, DC: AACC Press, 2002.

Horwitz M. Basic concepts in medical genetics. A student's survival guide. New York: The McGraw-Hill Companies, Inc., 2000.

715. Strachan T, Reed AP, eds. Human molecular genetics 3. London: Garland Science Taylor & Francis Group, 2004.

716–719. Shulenin S, Nogee LM, Annilo T, Wert SE, Whitsett JA, Dean M. ABCA3 gene mutations in newborns with fetal surfactant deficiency. N Engl J Med 2004;350(13):1296–1303.

720. Plant N. Molecular toxicology. New York: BIOS Scientific Publishers, Taylor & Francis Group, 2003.

721–723. Yue L, Saikawa Y, Ota K, Tanaka M, Nishimura R, Uehara T, et al. A functional single-nucleotide polymorphism in the human cytidine deaminase gene contributing to ara-C sensitivity. Pharmacogenetics 2003;13(1):29–38.

724. Parkinson A. Biotranformation of xenobiotics. In: Klaassen CD, ed. Casarett & Doull's toxicology. The basic science of poisons, 6th ed. New York: McGraw-Hill Medical Publishing Division, 2001.

725. Horwitz M. Basic concepts in medical genetics. A student's survival guide. New York: The McGraw-Hill Companies, Inc., 2000.

726. Strachen T, Reed AP, eds. Human molecular genetics 3. London: Garland Science Taylor & Francis Group, 2004.

727. Salto-Tellez M, Lee SC, Chiu LL, Lee CK, Yong MC, Koay ESC. Microsatellite instability in colorectal cancer: Considerations for molecular diagnosis and high-throughput screening of archival tissues. Clin Chem 2004;50(6):1082–1084.

728. Williams BJ. Comparative genomic hybridization. In: Cotter FE, ed. Molecular diagnosis of cancer. Totowa, NJ: Humana Press, 1996.

729. Kempski HM. The characterization of chromosomal abnormalities using fluorescence in situ hybridization procedures. In: Cotter FE, ed. Molecular diagnosis of cancer. Totowa, NJ: Humana Press, 1996.

730. Horwitz M. Basic concepts in medical genetics. A student's survival guide. New York: The McGraw-Hill Companies, Inc., 2000.

731–734. Bai R-K, Wong L-JC. Detection and quantification of heteroplasmic mutant mitochondrial DNA by real-time amplification refractory mutation system quantitative PCR analysis: A single-step approach. Clin Chem 2004;50(6): 996–1001.

735. Horwitz M. Basic concepts in medical genetics. A student's survival guide. New York: The McGraw-Hill Companies, Inc., 2000.

736–741. Bullinger L, Döhner K, Bair E, Fröhling S, Schlenk RF, Tibshirani R, et al. Use of gene-expression profiling to identify prognostic subclasses in adult myeloid leukemia. N Eng J Med 2004;350(16):1605–1616.

Valk PJM, VerVerhaak RGW, Beijen A, Erpelinck CAJ, van Waalwijk SB, van Doorn-Khosrovani JM, et al. Prognostically useful gene-expression profiles in acute myeloid leukemia. N Engl J Med 2004; 350(16):1617–1628.

Grimwade D, Haferlack T. Gene expression profiling in acute myeloid leukemia. N Engl J Med 2004;350(15):1676–1678.

Liu ET, Karaturi KR. Micorarrays and clinical investigations. N Engl J Med 2004;350(15):1595–1597.

742. Horwitz M. Basic concepts in medical genetics. A student's survival guide. New York: The McGraw-Hill Companies, Inc., 2000.

743. Shetty A, Forbes A. Pharmacogenomics of response to anti-tumor necrosis factor therapy in patients with Crohn's disease. Am J Pharmacogenomics 2002;2(4):215–221.

744. Cradic KW, Wells JE, Allen L, Kruckeberg KE, Singh RJ, Grebe SKG. Substitution of 3′ phosphate cap with a carbon-based blocker reduces the possibility of fluorescence resonance energy transfer probe failure in real-time PCR assays. Clin Chem 2004;50(6):1080–1082.

745. Horwitz M. Basic concepts in medical genetics. A student's survival guide. New York: The McGraw-Hill Companies, Inc., 2000.

746–749. Diverio D, Luciano A, Riccioni R, Coco FL, Biondi A. Analysis of the PML/RAR-α fusion gene in acute promyelocytic leukemia by reverse-transcription polymerase chain reaction. Technical recommendations, advantages, and pitfalls. In: Cotter FE, ed. Molecular diagnosis of cancer. Totowa, NJ: Humana Press, 1996.

750. Senden IP, de Groot CJM, Steegers EAP, Bertina RM, Swinkels DW. Preeclampsia and the C282Y mutation in the hemochromatosis (HFE) gene. Clin Chem 2004;50(5):973–974.